Tufts
6th August 2008

To Jo-Ann

Many thanks for all your
support with this project!

With best wishes

Kit

IMAGINING THE ELEPHANT

A Biography of Allan MacLeod Cormack

IMAGINING THE ELEPHANT

A Biography of Allan MacLeod Cormack

CHRISTOPHER L VAUGHAN
University of Cape Town, South Africa

Imperial College Press

ICP

Published by

Imperial College Press
57 Shelton Street
Covent Garden
London WC2H 9HE

Distributed by

World Scientific Publishing Co. Pte. Ltd.
5 Toh Tuck Link, Singapore 596224
USA office: 27 Warren Street, Suite 401-402, Hackensack, NJ 07601
UK office: 57 Shelton Street, Covent Garden, London WC2H 9HE

British Library Cataloguing-in-Publication Data
A catalogue record for this book is available from the British Library.

IMAGINING THE ELEPHANT
A Biography of Allan MacLeod Cormack

ISBN-13 978-1-86094-988-3
ISBN-10 1-86094-988-6

Printed in Singapore by Mainland Press Pte Ltd

For Allan's Family

Contents

Foreword

I would like to thank Professor Kit Vaughan most sincerely for affording me the great honour of inviting me to write this foreword, especially as I am the least capable of doing it justice.

I also thank him for the excellent research he has undertaken to write this biography in memory of my younger brother Allan MacLeod Cormack and to make it so interesting and instructive. Kit personally travelled around America, England, Scotland and South Africa to speak to Allan's family, friends and colleagues to ensure every detail was correct.

Maybe advanced years afforded me this honour. Old age can sometimes be uncomfortable. The body tells you Old Father Time is rapidly advancing. However, if you go out to meet him, remembering all the good things life has brought, there is no need to feel despondent.

Now, I come back to the birth of my younger brother, Allan, on 23rd February 1924. I was six at the time and was so proud to be sister to this wonderful baby. This pride continued all through our lives together and remembering those years is one of the 'perks' of old age.

I was fortunate to get to know Allan's wife, Barbara, and their three children, Margaret, Jean and Robert. What a talented family they are! We enjoyed their company when visiting America: they could not have been kinder to our family. We were so pleased that they (except Robert) attended our daughter's wedding in Cape Town in 1981.

Here I wish to pay tribute to our Mother who was left a widow in her early 50s when our father died in 1936. She did her utmost to point her three children in the right direction.

Allan had a very astute mind, a wonderful imagination, was full of fun and could be a great tease. When younger this sometimes made me furious but having two brothers I learned to cope with it very easily in later years.

His love of music remained with him all his days, although he didn't actually play an instrument except for a short time learning the piano. I know that Barbara and her daughter Jean still have season tickets for the Boston Symphony Orchestra.

Allan loved mountaineering as did our son George. He and Allan had a great relationship. According to Allan's family, he and George 'reclimbed' every peak in the Western Cape when George visited the Cormacks in America.

It is most gratifying to know that the proceeds from the sale of this book are to be donated to the Allan Cormack Book Fund at the University of Cape Town. This fund assists needy students wishing to study mathematics and physics in the Faculty of Science. All good wishes to the students who benefit from this fund. May they work hard, play fairly and set their sights on being yet another Nobel Laureate for South Africa.

Amelia McLeod Read
Cape Town
December 2007

Preface

I think I am an Elephant,
Behind another Elephant
Behind *another* Elephant who isn't really there ...

AA Milne (1927)

In December 2007, we celebrated the 40th anniversary of the world's first heart transplant, performed by Dr Christiaan Barnard and his team at Groote Schuur Hospital. That event on 3 December 1967 focused international attention on Cape Town and led directly to the recognition that South Africans were capable of conducting world-class research in medical science. As important as Barnard's contribution was, however, I believe that an even more momentous breakthrough in medical science took place ten years earlier at the University of Cape Town. It is the story of computer assisted tomography (or CAT).

In August 1957, Allan MacLeod Cormack returned to the University of Cape Town (UCT) from a year's sabbatical at Harvard University. Although he had accepted a position as an Assistant Professor of Physics at Tufts University in Boston, the Council of UCT required him to return to Cape Town for three months to fulfill his teaching obligations to the university. On Thursday 5 September, in the basement of the physics department, Allan conducted the experiment that would later lead to a seminal publication in the *Journal of Applied Physics* in 1963. This research, which had been initiated during the first six months of 1956 when Allan worked with Dr James Muir Grieve in the radiology department at Groote Schuur Hospital, ultimately led to the award of the Nobel Prize for Medicine in 1979, an honour that he shared with Godfrey Hounsfield.

When Allan died on 7 May 1998 I approached his family, both in South Africa and in the USA, with a view to creating a research group at the University of Cape Town that would honour his contributions to science. They readily agreed to my request and in May 2000 we launched the Medical Imaging Research Unit, which has been inspired by the idea that it is possible to conduct world-class research here on the southern tip of the African continent.

This biography on Allan Cormack had its genesis in 2003 when I spent a sabbatical year in Ireland as a Walton Fellow in the Department of Electrical Engineering at University College Dublin. Ernest Walton was Ireland's first (and so far only) Nobel Laureate in the natural sciences, having shared the physics prize with John Cockroft in 1951 for "their pioneer work on the transmutation of atomic nuclei by artificially accelerated atomic particles". In celebration of the centenary of Walton's birth, Vincent McBrierty, an Emeritus Professor of Polymer Physics at Trinity College Dublin, published a biography on Ernest Walton in 2003. That got me thinking. Maybe I could do the same thing for Allan Cormack.

In mid-2004 I approached the family with my idea and again they responded positively. The past three years have been an odyssey for me, as I have travelled from Cape Town to England, visiting Cambridge University, where Allan studied from 1947–1950, to Scotland, where his forebears all originated, to Boston in the USA, where he spent the last 40 years of his life. As you will see from the chapter titles, I have extended this analogy of a journey to the description of Allan's life. It begins in the far north of Scotland, at John O'Groats, stops off at Jo'burg, Table Mountain, Cambridge, Cape Town, Boston and Stockholm, and finishes in Massachusetts.

One of the challenges for any author who completes a book is to select a title that captures the essence of the subject. My challenge was no different. The primary source for *Imagining the Elephant* comes from the famous Indian legend, captured in the poem by John Godfey Saxe (1816–1887):

> It was six men of Indostan,
> To learning much inclined,
> Who went to see the Elephant
> (Though all of them were blind),
> That each by observation
> Might satisfy his mind.

Two scientists, in writing about computer tomography, have also alluded to this legend. The first was the late Paul Lauterbur, with whom I exchanged e-mail messages in October 2004. He won the Nobel Prize for Medicine in 2003 for his work in magnetic resonance imaging and he knew Allan Cormack well. He and Allan were on the organising committee for a workshop entitled *Techniques of Three-Dimensional Reconstruction* that was held at the Brookhaven Laboratory on Long Island, New York in 1974. This is what Lauterbur said to me just over three years ago:

> Bob Marr and I toyed briefly with the idea of a follow-up meeting, until we realized that, unusually for scientific meetings, the first one had been so successful that another one was not needed – like the proverbial six blind men, we had all recognized the elephant together.

The other person was George Ellis, a distinguished cosmologist and applied mathematician at the University of Cape Town who wrote an essay on Allan Cormack for a recent book celebrating South Africa's Nobel Laureates. Here's what Ellis wrote:

> Consider the age-old example of the elephant. If all you had ever seen of an elephant was a photograph taken from one viewpoint, you would not understand it very well. A series of pictures from a whole variety of vantage points would let you know much more about it – indeed you can build up a 3D image in your mind from an array of 2D images.

This is the essence of tomography, which is captured in the whimsy of AA Milne's poem about Christopher Robin in *Now We Are Six*, and quoted at the beginning of this preface. A CAT scanner allows us to uncover the identity of objects that would otherwise be hidden in a traditional 2D X-ray image. When confronted with this challenge 50 years ago, Allan started to think about the problem *ab initio*. It did not take him long to realise the problem was a mathematical one and so he set about devising a suitable algorithm for solving the problem.

There's another connection between Allan Cormack and the elephant. In December 1945 he and his friends from UCT hiked and climbed in the Cederberg, a mountain range about two hours' drive north of Cape Town. They came across an ancient rock painting, estimated to be two thousand years old, illustrating the Khoisan bushmen and elephants. I know this because I discovered a black and white photograph of the painting in Allan's personal album (see Figure 2.9). While the bushmen and elephants can no longer be found in the Cederberg, the stunning painting endures and a photo of it has been used to illustrate the back cover of this book.

I twice heard Allan speak. The first occasion was on Monday 16 March 1981 when he delivered a public lecture at UCT on 'Computer Tomography: Past and Future Portent'. His self-deprecating humour was in evidence, with the audience particularly enjoying his recollection of the lack of interest shown in his original publications and his anecdote of a CAT scanner being used in the search for a skier buried under an avalanche!

The second occasion was on Saturday 16 April 1994 when Allan was a keynote speaker at the Thirteenth Southern Biomedical Engineering Conference in Washington DC, and I was a Professor at the University of Virginia in nearby Charlottesville. He spoke on 'CT Scanning up to the Appearance of the EMI scanner' and again I was struck by his *joie de vivre* and genuine love of science. So, although I was never formally introduced to Allan MacLeod Cormack, I feel that I have come to know him well during these past three-and-a-half years.

I have written the text in a style and at a level that I hope will be accessible to all readers who have an interest in biography, and not just those scientists who may wonder how it was that a physicist, who solved a mathematical algorithm, came to win the Nobel Prize in Medicine. I trust you will enjoy reading this story about a family man who was often described as a 'modest genius'.

Christopher Leonard (Kit) Vaughan
Cape Town
January 2008

Acknowledgments

When I embarked on this journey in mid-2004, I had little idea what the demands would be. Now that I have reached my destination, it is clear that I could not have completed the journey without the advice and enthusiasm of many people. It is a pleasure for me to acknowledge their contributions here.

There were four special people without whom the project could not have happened. These were: Amy Read, Allan's older sister, who was my strongest supporter and whose memory of events and people from over 70 years ago was really quite astonishing; Jean Cormack, Allan's younger daughter, who, shortly after her father's death in May 1998, had the foresight to ask his old friends to write down their recollections of shared experiences; Gareth Vaughan, my son, who tracked down many of Allan's early publications and who edited every chapter; and Margaret Cormack, Allan's older daughter, who provided family details and insights about her father, and who also edited every chapter.

Besides Amy, Jean and Margaret, there were 17 other people whom I interviewed face-to-face. These were: Barbara Cormack, Allan's wife, who generously gave me copies of letters she wrote to her parents from Cambridge University in 1948–49; Robert Cormack, Allan's son, who co-authored a paper with his father while he was still an undergraduate at Harvard and has pursued a career in medical physics; Len Read, Allan's brother-in-law, who provided details of life at the University of Cape Town in the early 1950s; Dorothy and Bob Seavey, Allan's sister- and brother-in-law, who provided perspectives on life at the family retreat in Alton, New Hampshire; Michael de Lisle, one of Allan's climbing companions from 1945–46, who loaned me his logbook from that time; Walter Powrie, Allan's oldest friend from his high school days, who shared Allan's passion for climbing; Howard Phillips, who gave me invaluable advice about archives research, and whose book *The University of Cape Town: 1918–1948* was a source for some of the anecdotes about Allan's undergraduate years; Andy Koehler, director of the Harvard Cyclotron Laboratory, who provided Allan with friendship and intellectual stimulation for over 30 years; Todd Quinto, Professor of Mathematics at Tufts University, who collaborated with Allan in extending the applications of Radon's transform; Ken Astill, Professor of Mechanical Engineering at Tufts, who remembered Allan as a man with a keen sense of humour who enjoyed being politically incorrect; Kathryn McCarthy, Professor of Physics at Tufts and, when she was Provost, one of Allan's most important supporters; Steve Webb, who shared with me his correspondence with Allan, and whose book *From the Watching of Shadows* provided key insights into the history of computed tomography; Godfrey Stafford, who overlapped with Allan at UCT and Cambridge and who generously hosted me in his home at Oxford; John Wanklyn, who first met Allan as an undergraduate at UCT in 1942 and was still corresponding with him over 50 years later; Aaron Klug, who studied physics with Allan at UCT and who would also win a Nobel Prize; and Robin Cherry, Professor of Physics at UCT and Allan's first postgraduate student in the early 1950s.

There were two people whom I interviewed by telephone. These were: Bernard Gordon, who visited Allan in October 1975 to seek his advice on computed tomography and who later developed the instant imaging CAT scanner; and Burton Hallowell, President of Tufts University, who in February 1976 explored the possibility of nominating Allan for the Nobel Prize.

One of the benefits of the modern era has been the internet which enabled me to track down individuals who had information about Allan. The following people answered my questions via e-mail: George Lindsey, one of Allan's fellow Nuclear Nit-Wits, who provided an essay on life at Cambridge University and the Cavendish Laboratory in the post-war years; Colin 'Boot' Butler, a climbing companion of Allan's who also shared his commitment to the teaching of physics; Brenton Stearns, a fellow Assistant Professor of Physics with Allan at Tufts in the late 1950s; David Hennage, an undergraduate physics student, who assisted Allan with his seminal research in the early 1960s; David Ayres, a particle physicist with whom Allan collaborated at Berkeley in 1966–67; Rodney Brooks, who did research with Allan on proton tomography and who also published a review article that drew attention to Allan's research on computer tomography;

Michael Goitein, who developed his own computer tomography algorithm and tested it with Allan's data, and who later pursued a career in the physics of radiation therapy at Harvard; Bob Marr, who invited Allan to serve on the organising committee for a workshop entitled *Techniques of Three-Dimensional Reconstruction* in 1974; Paul Lauterbur, whom Allan first met at the workshop and who would later win the Nobel Prize in Medicine for his research on magnetic resonance imaging; Matt Howard, a distinguished and creative neurosurgeon, who, as an undergraduate physics major at Tufts, was inspired by Allan's Nobel Prize; Robert Lewitt, who, as a post-doctoral research fellow in 1977, had the temerity to challenge the mathematics in one of Allan's publications; Marcella Tanona in the Provost's Office at Tufts, who tracked down information on the award of the Hosea Ballou Medal to Allan in 1978; Rose Mary Logue, of University College Dublin, who provided me with the history of *Cormac Mac Airt*, the most renowned of the ancient kings of Ireland; Brian Warner, Professor of Astronomy at UCT, who supplied key details about Allan's first journal publication in 1945; Maarten de Wit, Professor of Geology at UCT, who provided background information for Allan's second journal publication in 1947; and Cornelis Plug, Professor of Psychology at the University of South Africa, who supplied me with the details of the Jubilee Science Congress held in Cape Town in 1952.

There were then those who gave me something that was critical to the biography. These included: John Juritz, a fellow physicist who studied and worked with Allan at UCT in the 1940s and 1950s, who provided me with old letters and newspaper cuttings; Dick Wilson of Harvard University and a collaborator of Allan's in the 1950s and 1960s, who sent me a copy of *A Brief History of the Harvard Cyclotrons*; Torgny Greitz, one of Allan's champions who introduced him at the Nobel award ceremony in Stockholm in December 1979, who provided me with the only photographs ever taken of Allan and fellow Laureate Godfrey Hounsfield; Annika Ekdahl of the Nobel Foundation, who sent me material from the Nobel week in December 1979 and gave me permission to reproduce Allan's lecture (see Appendix B); Paul Lagrange, archivist at Rondebosch Boys' High School, who gave me a copy of the book *Rondebosch Boys' High and Preparatory Schools 1897–1997*; Anne Dunnett of Wick and Margaret MacKay of Edinburgh, who tracked down details of Allan's family all the way back to the late 18th century; Susan Lapides, a photographer who dug out the old negatives of photos she had taken of Allan in October 1979 and gave me permission to reproduce them; Fiona Colbert who gave me the texts *St John's College,*

Cambridge and *Portrait of a College*, and also provided me with archival material on Allan and other Johnians; Gerald Friedland, an emeritus professor of radiology at Stanford, who gave me copies of his correspondence with Allan as well as reprints of papers he had written about Allan; and Leah Ellis of Quest Productions, who sent me a videotape interview and transcript done with Allan for the TV programme called *Naked to the Bone.*

There were others who provided me with a key service. These included: Cathy Hole of UCT, who transcribed all the interviews; Nazir Karbanee and Charles Harris of UCT, who created the engineering drawings and built the replica of Allan's original CAT scanner (Figure 5.4); Kelvin Fagan of the Cavendish Laboratory in Cambridge, who scanned photographs of equipment and people from the late 1940s; Rich McManus of the National Institutes of Health, who scanned a photograph of Allan and other pioneers of computer tomography; Rik De Decker of UCT, who tracked down a 1946 description of *Cormack's Way*, a climbing route up Table Mountain opened by Allan; Natsumi Tanimura, a Japanese student studying at UCT, who translated the article about Allan that appeared in a teenager's magazine called *Shonen Jump*; Annelie O'Hagan, an artist in Cape Town who reproduced the artwork from the *Shonen Jump* article; Michael Wyeth of Imago-Visual in Cape Town, who designed the cover to this book; and Barbara Elion of One Life Media in Cape Town, who assembled the index.

Not surprisingly, I discovered that a biography is dependent on archival material normally housed in libraries. At Rondebsoch Boys' High School, I was assisted by Peggy Quayle. At the University of Cape Town, those who helped me included Lesley Hart, Stephen Harandien, Sipho Masha and Emily Krige. After Allan died in 1998, his family donated all his papers to the archives section of Tisch Library at Tufts University and it was here that I discovered much of the source material for this book. I spent two productive periods in the archives, in October 2004 and April 2006, and my time there was facilitated by Susanne Belovari, Jo-Ann Michalak and Anne Sauer.

The University of Cape Town generously provided me with financial support and sabbatical leave to work on the biography in 2006. I also acknowledge my editors at Imperial College Press, Lance Sucharov and Katie Lydon, who have been unflagging supporters of this project, despite numerous delays. Finally, I am pleased to recognise my wife Joan who, for over a year, endured my 5:00 am start to the day when I would go and "talk to Allan". Without her understanding, this book would not have seen the light of day!

Chapter 1

From John O'Groats to Jo'burg

> What better place to start than with your own family and how
> they have been formed by history and, in their own small way,
> have contributed to it?
>
> Allan Cormack[1]

Allan Cormack's forebears all came from Caithness, the northernmost extent of the Scottish mainland (Figure 1.1). Edging a rugged treeless landscape, the Caithness coastline consists of sheer cliffs interspersed with beautiful beaches and two harbours – Wick and Thurso – at the mouths of the principal rivers. Wick, which takes its name from the first three letters of 'Viking', was the birthplace of Cormack's parents, George Cormack and Amelia MacLeod. The area has a rich history, stretching back over five thousand years, and it held a particular fascination for Allan.

Neolithic farmers, probably from mainland Europe, settled the area around Caithness in about 4500 BC, just 1500 years after the end of the Ice Age, during which Britain was covered with a blanket of ice. These farmers left behind many impressive structures, including the village at Skara Brae and a chambered tomb called Maes Howe, located just 50 miles north of Wick on the Orkney Islands. Together with the burial tombs such as New Grange just north of Dublin in Ireland, these ancient edifices are said to rival the Egyptian structures of the same era in both ingenuity and finish. The next inhabitants to settle in Caithness were the Beaker people who ushered in the Bronze Age, a period lasting from 2000 to 700 BC.[2]

The Celts, called Keltoi by the Greeks who regarded them as barbarians, arrived around 800 BC. They not only brought the Iron Age to all

1

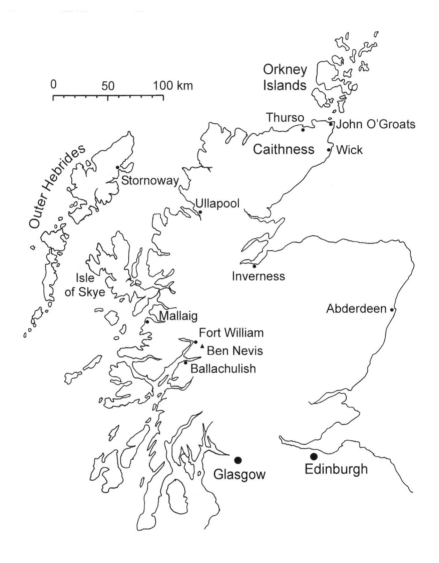

Figure 1.1   Map of Scotland

of Britain and Ireland, but their languages also became the basis for all
languages in pre-Roman Britain. Their fortified houses, known as *brochs*,
many fine examples of which can be found in Caithness and the Orkney and
Shetland Islands, have been described as "the most ingenious and impressive military works of prehistoric man in Western Europe".[3] The master

plan incorporated a circular tower, with double walls, a space between the inner and outer walls, and stone landings connected by a winding staircase. Each broch required almost 5,000 tons of stone and was constructed without mortar or lime. With just a single entrance, defended by guardrooms and a great inner door, the tower was impregnable.[4]

Although the Romans annexed and controlled Britain between the first and fifth centuries AD, they never extended their area of influence further north than the Firths of Forth and Clyde, the line between which joins the modern cities of Edinburgh and Glasgow (Figure 1.1). However, the Romans did leave behind the names of two Celtic tribes: the *Scotti*, the fierce raiders of their north-west outposts, which ultimately led the area north of England to be called Scotland; and the *Caledonii*, who later became known as the Picts because they tattooed their foreheads (*pictus* is the Latin word for *painted*). When the Romans had departed Britain, the Picts began raiding south, with the result that the Britons there invited the Angles from Denmark and the Saxons from north-west Germany to assist with their defence. Many stayed and their influence on the English language has been far-reaching.[5]

In the 8th and 9th centuries AD the Vikings extended their territory: west from Scandinavia to Iceland, Greenland and, very briefly, North America; and south to the Mediterranean Sea, encompassing Rome and Constantinople. Because of their geographic proximity the Norwegian Vikings occupied the coastal regions of Caithness as well as the Orkney and Shetland Islands. The history of Caithness, from the mid-9th to the mid-13th centuries, is recounted in the Icelandic record known as the *Orkneyinga Saga*.[6] The Norse settlers had a profound effect on Pictish land, leaving behind place names such as Thurso (from the Viking god Thor), Lybster and Stemster (*stedr* is the Old Norse word for *place*) and, as mentioned previously, Wick. The Earls of Orkney and Caithness, first under the patronage of Norwegian kings, controlled the area up until about 1300, and thereafter they ruled with the backing of Scottish kings. By the middle of the 15th century both earldoms were in the hands of the St Clair family (also known as Sinclair) and in 1472 Scotland formally annexed the Orkney and Shetland Islands.[7]

At the end of the 18th century one of Allan Cormack's distant relatives, Sir John Sinclair of Ulbster, compiled *The Statistical Account of Scotland* and, alongside many other regions, described the Caithness area.[8] He was, apparently, a man of "enormous energy and unbounded self-conceit" and was responsible for introducing the word *statistics* to the English language.[9]

He based his account on reports submitted by the Presbyterian ministers of each parish – tardy clergymen were admonished with a sharp reminder written in red ink! For Caithness, there were three recurring themes: the state of agriculture; emigration; and the consumption of alcohol.

Because the majority of farms had absentee landowners who were not prepared to offer long leases to their tenants, the state of agriculture was deplorable. There were few opportunities for enterprising young men and many chose to emigrate: south, to the industrial mills or the military; and west, to join the Hudson Bay Company and help to settle Canada. Almost without exception, the ministers were distraught at the number of inns and pubs in their parishes as well as the volume of liquor consumed. When Sinclair's account was updated in 1845, the minister of the Wick parish described the status of mental health thus: "Maniacs are very rare. Idiots and fatuous people are remarkably common"![10]

Despite these considerable impediments, the North Sea provided the people of Caithness with a generous bounty in the form of enormous shoals of herring. By the middle of the 18th century political and other barriers to fishing had been removed, and in 1767 the first boats set sail from Staxigoe, just two miles to the north of Wick.[11] Before the turn of the century there were more than 150 boats plying their trade off the Caithness coast but, with the lack of harbours and port infrastructure, expansion became imperative. The British Fisheries Society supported the construction of a new harbour and a new village on the south bank of the river Wick. By 1807 Pulteneytown had been built, complete with terraced houses and ship yards for the herring fishing industry. The harbour was completed by 1811 and within five years the town had reached its target of 1,000 residents.

The rise and fall of the fishing industry was central to the lives of Allan Cormack's family during this period. His great-great-grandfather, John Cormack (1790–1862), was a ship captain, having begun his career as a fisherman, and perished at sea at the age of 71 (Figure 1.2). He and his wife Elizabeth Thomson (1789–1872) resided at Martha Terrace, Pulteneytown for many years.

The permanent population of Caithness grew rapidly from 17,000 in 1802 to over 40,000 by the mid-1840s. Not surprisingly, there was a large itinerant population during the fishing season and, much to the consternation of the Presbyterian clergymen, a commensurate increase in the consumption of alcohol and its attendant problems. This resulted in Wick being designated a 'dry' town, a status that was maintained until the First World War. Notwithstanding this setback, the number of registered fishing

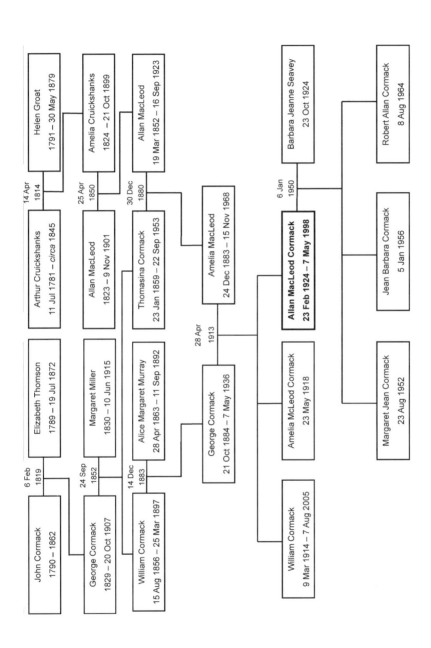

Figure 1.2 Allan Cormack's family tree.

vessels in Wick had grown to 1,120 by 1862 and the harbour had reached its capacity.

Allan Cormack's maternal great-grandfather, Allan MacLeod (1823–1901), was an Assistant Harbour Master in 1863 when a new breakwater was started under the supervision of Thomas Stevenson. Thomas was the son of the famous family of lighthouse builders and lived with his wife and son Robert Louis Stevenson in Harbour Terrace, Pulteneytown for the duration of the project. Living nearby in the same terrace was Allan MacLeod, his wife Amelia Cruickshanks (1824–1899) and their son Allan (1852–1923). In later years Robert described Wick as "The baldest of God's towns on the bleakest of God's bays".[12] While this statement captures the desolate Caithness coastline, his comments were probably influenced by his father's inability to complete the breakwater. By 1873 the project lay in ruins and, despite the significant costs already incurred, was abandoned.

Allan Cormack's paternal great-grandfather, George Cormack (1829–1907), began his career as a fisherman, captaining a boat called Maggie,[13] and in the last decades of the 19th century set up a fish curing business, G Cormack & Son, in Pulteneytown. He and his wife, Margaret Miller (1830–1915), lived at 5 Harbour Quay, overlooking the Wick harbour. This stone-faced building incorporated not only their living quarters but also the business office and a yard for processing the herring. As indicated by the company name, George was assisted by his son William (1856–1897) and his son-in-law Allan MacLeod (1852–1923) who had married his daughter Thomasina (1859–1953). Portraits of three of Allan Cormack's great-grandparents (Figure 1.3) and three of his grandparents (Figure 1.4) were recorded by the renowned plumber-turned-photographer from Wick, Alexander Johnston.

While improvements in technology, such as the introduction of steam power, led to a reduction in the registered fleet to 600 boats, the years from the turn of the century to the beginning of World War I in 1914 were marked by prosperity. During this period there were 90 companies participating in the fishing industry in Wick. The war, however, marked the beginning of the decline in the industry. Following the economic depression of the 1920s and 1930s and the disastrous impact of the Second World War, the herring fishing industry finally came to an end by the mid-1950s.

One of Allan Cormack's maternal great-great-grandmothers was Helen Groat (1791–1879) who was married to Arthur Cruickshanks (1781– c. 1845). Helen's father was John Groat, a farmer, her mother was Elizabeth Sinclair and she is said to have been the "last of the Groats".[14] She was a

Figure 1.3   Allan Cormack's great-grandparents: George and Margaret Cormack (left); and Allan MacLeod (right).

descendant of John O'Groats, after whom the famous northernmost settlement was named(Figure 1.1). Although various stories regarding the origin of the name abound, the most plausible is that the Earl of Caithness in 1496 issued a charter to John Groat, son of Hugh Groat who was the Earl's steward: the Earl donated the land from which John established a ferry service to the Orkney Islands.

The origin of the name *Cormack*, which is said to mean 'born in a cart', may be traced to Ireland. Cormac Mac Airt, the son of Art, is perhaps the most renowned of the ancient kings of Ireland, and ruled at Tara for forty years from 227 AD.[15] Described in the Annals of Clonmacnoise as "absolutely the best king that ever reigned in Ireland before himself", he was respected for his wise, true and generous judgments, and was the author of *Teagusc Na Righ*, which was designed to preserve the manners and morals of the kingdom.[16] He is reputed to have fought many battles, subduing the provinces of Ulster and Connacht and waging a drawn-out campaign against Munster in the south of the country.

Cormac also led the first recorded raids on Roman Britain and Pictish Scotland, and apparently fell in love with a beautiful woman called Cearnait, daughter of the King of the Picts.[17] When he died in 266 AD,

Figure 1.4 Allan Cormack's grandparents: William Cormack (left); and Thomasina and Allan MacLeod (right).

apparently from a salmon bone that stuck in his throat, Cormac was buried, according to his wishes, not at New Grange on the River Boyne (the traditional burial ground of Irish kings) but further downstream at Rosnaree, facing eastwards towards the rising sun. Although some experts have argued that he was a mythical figure, and indeed many Irish legends have been attached to his name, there is enough evidence to suggest that Cormac Mac Airt was an authentic historical figure.[18]

With all the Viking traffic between the Orkneys, Ireland and Iceland, the name Cormack obviously travelled around, and featured in an Icelandic saga, *Kormak the Skald*. One of its earliest recorded appearances in northern Scotland has been dated to the mid-16th century. As published in the *Origines Parochiales Scotiae*, Mary, Queen of Scots, granted remission to 50 men from Caithness – including Cormack, Cruikshank and Groat – for their "treasonable taking and holding of the castle, house and place of Ackirgill".[19]

Allan Cormack's parents, George Cormack (1884–1936) and Amelia MacLeod (1883–1968), were first cousins (Figure 1.2). George's parents, William Cormack and Alice Murray, died relatively young, aged 40 and 29 respectively, so he was brought up by his paternal grandparents, George and Margaret Cormack (Figure 1.3). When the 1901 census was conducted,

Figure 1.5 Allan Cormack's parents: George Cormack (left); and Amelia MacLeod (right).

16-year-old George was registered as living at 5 Harbour Quay, Wick, and as working as a telegraphic messenger worker (Figure 1.5). In 1903 he emigrated to the British colony of Natal as a telegraphist and was stationed at Mooi River. This was the year after the Anglo-Boer War had ended and well before the Union of South Africa was established in 1910. On 30 December 1903 the elder Cormack wrote to his grandson, thanking him for the letter describing his safe arrival in Natal and providing some grandfatherly advice:

> It gives us pleasure to see that you are among friends and I hope that they will be seting [sic] you a good example. Above all things keep away from bad company and be shur [sic] and keep straight in all your dealings.[20]

In the same census of 1901, 17-year-old Amelia MacLeod was registered as living at 15 Smith Terrace, Pulteneytown with her parents Allan and Thomasina MacLeod, and employed as a pupil teacher (Figure 1.5). In this capacity she taught at the local school in Wick under the supervision of a qualified teacher. Shortly thereafter Amelia was sent to Moray House in Edinburgh where she received formal training as a teacher. On the first of these trips her grandfather George Cormack travelled down south with her and they spent a night at a local hotel. It was the first time the young Caithness lass had ever seen an electric light switch – they still had gas lamps in Wick – and she kept turning the switch on and off.[21] Amelia and her cousin George were by all accounts good correspondents and kept in regular contact. George, employed in the engineering department of the Post Office, trained as a technician in Pietermaritzburg, the capital city of Natal. Then in early 1913, after a decade apart, Amelia set sail for South Africa, disembarking at Durban harbour on the morning of 28 April. George was there to meet her and that afternoon, in the Presbyterian Church, they were married (Figure 1.2).

South Africa is a land of magnificent and breathtaking scenery which includes deep valleys, towering mountains and broad plateaus. The Cape mountain region, with its extraordinary floral diversity, stretches from the Namib Desert in the west to the coastal region of Natal in the east. The Drakensberg mountain range, rising 11,000 feet above sea level, forms the eastern escarpment, separating the coast from the inland plateau, also known as the Highveld (Figure 1.6). The geological formations on the Highveld yielded a bounty of natural resources – diamonds, gold, platinum and coal – which were discovered in the last decades of the 19th century. For the young couple from Caithness, the country appeared to offer unlimited possibilities.

After the British victory in the Anglo-Boer War in 1902, there was an opportunity to place the four colonies – the Cape, Natal, Orange Free State and the Transvaal – on an equal footing. The franchise rights were held by the white authorities and by the time the Union of South Africa was formed on 31 May 1910, the only province with a non-racial franchise was the Cape.[22] Black South Africans were barred from membership of parliament, despite constituting almost 80 percent of the country's 6 million inhabitants. The South African Party, an amalgamation of Afrikaner parties, held power under the premiership of General Louis Botha, and it did not take them long to entrench white power with a raft of new laws. Among these were the pass laws – which required black citizens to carry identity documents at all times – and the Land Act of 1913 which set aside 90 percent of the country's land for white ownership. Before this act was promulgated, the African National Congress (ANC) was formed on 8 January 1912, bringing together black citizens, including an educated elite and rural peasants. It would be over 80 years before the ANC was able to claim its rightful place at the table of South African politics.

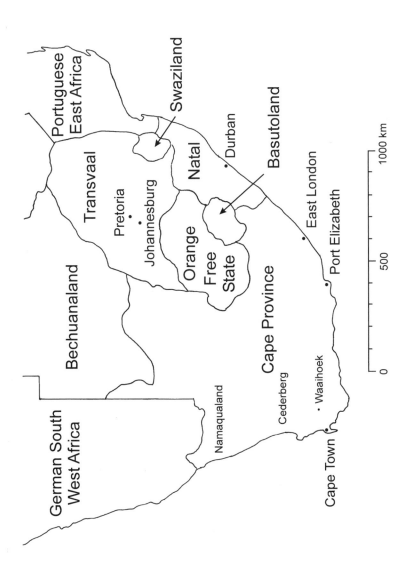

Figure 1.6   Map of South Africa at the time of Union in 1910.

George and Amelia Cormack did not stay in Natal, as George was transferred to the Post Office at Johannesburg in the Transvaal. In the short
period after the discovery of gold on the Witwatersrand in 1886, the town
had expanded rapidly. Such was the extent of the mining industry's growth,
based particularly on the extraordinary reserves of gold, that in time Johannesburg would become the financial nerve centre of the country. Key
to this growth were the telegraph and telephone, which were operated and
controlled by the South African Post Office. George Cormack, with his
training and experience as an engineering technician, was therefore well
placed to make an important contribution to the burgeoning telecommunications industry in the country's heartland.

The Cormack's first child William, named in the Scottish tradition after
his paternal grandfather, was born on 9 March 1914 in Johannesburg and
news of the happy event was relayed via letter to the family in Caithness.
Meanwhile, in Europe the storm clouds were gathering: the assassination
of Archduke Francis Ferdinand in Sarajevo on 28 June 1914 triggered the
outbreak of World War I.[23] South Africa, as part of the British Empire,
joined the Allied war effort. Jan Christiaan Smuts, a Cambridge-educated
lawyer who had fought against the British in the Anglo-Boer War, was
Minister of Defence in the first Union cabinet. During World War I he
excelled as field general in the German East Africa campaign, served in the
Imperial War Cabinet led by David Lloyd George and assisted with the
creation of the Royal Air Force.[24]

Just as Johannesburg and the mines of the Witwatersrand depended on
telecommunications, so too did the war effort. Although George Cormack
was not drafted into the military, he was nevertheless called upon to assist
and contributed through his engineering skills. This required him to travel
through to Pretoria, some 40 miles north of Johannesburg, and, on occasion,
to spend time away from his wife and young son. As the war in Europe was
drawing to a close, a daughter was born: Amelia McLeod Cormack entered
the world on 23 May 1918. Later the following year, with the cessation of
hostilities and the re-opening of shipping lines, Mrs Amelia Cormack and
her two children visited Scotland, where they stayed with her parents Allan
and Thomasina MacLeod at 54 Argyle Square, Wick. It was a joyous time
for the grandparents, enabling them to meet their grandchildren William
and Amelia, or Bill and Amy as they came to be known. This was the first
of numerous pilgrimages that the South African branch of the Cormack
family would make to their Caithness roots.

The end of World War I brought a period of economic depression to the gold mining industry in South Africa. In 1920 the price of gold dropped by over 30 percent to less than £5 an ounce and mining executives, faced with rising costs, were confronted with the grim possibility of closing mines and laying off tens of thousands of miners.[25] The Chamber of Mines, headquartered in Johannesburg, planned to reduce labour costs by removing job reservation (a policy which set aside skilled jobs for white miners) and increasing the proportion of black workers, who were paid at a lower rate. These actions, which white miners saw as a threat to their livelihood, led to widespread strikes the following year and in March 1922 there was a full-scale rebellion. Marauding gangs of armed white miners set up barricades and roadblocks in an effort to capture the city. They attacked the Johannesburg Post Office, where George Cormack was based, and also the city's power station.

Jan Smuts, who had become Prime Minister two years earlier following the death of Louis Botha, called in the military. Within two weeks the revolt had been crushed but the fallout was calamitous: over 250 people were killed, many of them policemen, and the cost of the damage to property was estimated to be tens of millions of pounds. Four of the ringleaders were condemned to death and marched to the gallows singing their anthem *The Red Flag*. Ironically the rebels had been supported by the Communist Party of South Africa, which later switched its allegiance to the African National Congress and the rights of black workers. Smuts himself became a victim of the fallout: he was heavily criticised for his severe handling of the rebellion and lost support among Afrikaans-speaking voters. He was subsequently defeated in the general election of 1924, relinquishing power to General JBM Hertzog and his National Party.[26]

Allan MacLeod Cormack was born 23 February 1924. He was named after his maternal grandfather, Allan MacLeod (Figure 1.4). The Cormack family was then living at 103 Becker Street, Bellevue, a middle class suburb just a few miles north-east of downtown Johannesburg.[27] His mother Amelia, in addition to her formal training as a teacher, was an accomplished watercolour painter and immersed herself in a variety of home crafts: knitting, crochet and dressmaking.[28] Young Amy, not quite six and homeschooled by her mother, was delighted with the arrival of a younger brother. She soon learned how to change his nappy and assisted her mother at bath time. Later that year, George and Amelia, together with their three children, boarded the train for the 800-mile journey south to the coastal city of Port Elizabeth. A family vacation at the beach was a fitting way to round off an auspicious year.

By the middle of 1926 George Cormack had accumulated sufficient leave from the Post Office for him to take an extended break. It was time for another pilgrimage by the Cormack family to their Caithness roots. Again they stayed with Thomasina, Amelia's mother, at Argyle Square. Thomasina's husband Allan had died in 1923 following a stroke so Allan Cormack never met the grandfather after whom he had been named. However, there was an opportunity for the Cormack children to get to know their Scottish relatives, particularly their mother's younger siblings, her sister Betty and brother John, neither of whom would ever marry. It was also a time to explore the harbour with their family's fishing heritage and to enjoy the beaches just north of Wick, at Reiss and John O'Groats (Figure 1.7). It would also be another 12 years before the Cormacks were able to return to Scotland although sadly it would be George's last visit to his ancestral home.

Figure 1.7 The Cormacks enjoy a family outing on the shore at John O'Groats, 1926.

Back in Johannesburg George was promoted and the family moved two streets farther south into a more spacious home at 16 Isipingo Street, Bellevue. That same year in a house diagonally opposite them, at 19 Isipingo Street, one of South Africa's most infamous murders took place.[29] Herman Charles Bosman, who would subsequently become one of the country's most celebrated writers, had been teaching at Groot Marico in the Karoo. He shot and killed – apparently accidentally – his step brother, David Russell, in his Bellevue home. Within months he was convicted of first degree murder and sentenced to death. Fortunately for Bosman, and indeed for those who would come to appreciate his considerable skills as a story teller, the death penalty was commuted to life imprisonment and less than four years later he walked away from Pretoria Central Prison, a free man.

Despite the ferment in the neighbourhood, the Cormack children (Figure 1.8) continued to enjoy the simple pleasures of domestic life. There was an addition to the family, a cocker spaniel that Allan named 'Rojo'. Then, on a crisp autumn morning in April 1930, there was a momentous event in the educational journey of young Allan: his first day at school. Following the British tradition, schools in South Africa had their own distinct uniform. The Observatory Junior School's colours were navy and white. The girls wore pinafores while the boys sported white shirts, striped ties, navy shorts and knee length socks. When stepping out they wore a smart navy blazer with the school's badge prominently displayed on the breast pocket. Allan was a gregarious child and soon established firm friendships with a number of his fellow pupils. One of his best friends in Ms Davies' first grade class was Tommy Butcher who, over sixty years later, would become the personal tennis coach to the Sultan of Brunei.[30]

On the political front, General Hertzog had introduced a segregation scheme in the shape of the Civilised Labour Policy, the Colour Bar Act of 1926, and a series of four Native Bills. The so-called "civilised" policy was in fact designed to marginalise black South Africans, and to prevent them from taking up positions in industries such as the railways. Blacks were repressed with the express purpose of uplifting indigent white citizens. The school syllabus was divided into Higher and Lower certificates for pupils along strictly racial lines: the state contributed £15 per white pupil but only £2 per black child.[31] Under the prevailing government policy the Cormack children all attended schools that were reserved for white pupils. Despite their best legislative efforts to favour white citizens, the government's popularity with the voters waned. This was a direct result of the Great Depression that threatened the world's economy in the early 1930s. In the Union of South Africa, Hertzog was forced into a coalition government with General Smuts, appointing him as his Deputy Prime Minister.[32]

Despite the widespread effects of the Depression, the Cormacks were able to purchase their first family motor car, a Triumph, imported to South Africa from Great Britain. For their summer vacation of 1931, George and

Figure 1.8   Family portrait of Bill, Allan and Amy Cormack, 1930.

17-year-old Bill drove the 400 miles over dirt roads to Doonside, Natal, while the other members of the family travelled by train. The following year George was transferred by the Post Office from Johannesburg to Cape Town. The family decided to give Rojo to their neighbour, Mrs Paton, rather than transport him to the Cape.[33] Unbeknown to his parents or siblings, 8-year-old Allan wrote a letter to Mrs Paton from Cape Town, demanding to know why she had taken his dog away! When Amelia discovered what had happened she was not amused and Allan was obliged to write a second letter, apologising for his behaviour. It was not long, however, before the family had acquired another dog, a fox terrier that they named Jock.

On the educational front, Allan was enrolled at Rondebosch Boys' Junior School, Amy attended Rustenberg Girls' High, while older brother Bill followed in his father's footsteps, and entered the first year electrical engineering class at the University of Cape Town (UCT). The family was renting a house called St Ronan's in Richmond Road, Mowbray. A friend of George Cormack gave him a pair of homing pigeons that were both supposed to be male.[34] It transpired they were a male and female pair and, not surprisingly, there were soon some offspring. A favourite game of Allan's was to take one of the pigeons to school in the morning and then to send it home with an amusing message attached to the bird's leg. Despite the obvious enjoyment they provided, George eventually decided to part with the pigeons.

The family had no sooner settled in Cape Town when two years later, in 1934, the Post Office again transferred George, this time to Port Elizabeth, 500 miles to the east (Figure 1.6). At Allan's insistence, Jock was not left behind (Figure 1.9). Bill, however, did remain in Cape Town to continue his studies at UCT, earning his degree in engineering at the end of 1935.[35] The Cormacks moved into rented accommodation, a third-floor flat strategically positioned on Military Road. From the balcony they looked up the hill towards Amy's high school, Collegiate, and down the road with a view of the harbour. Allan was enrolled at Grey Preparatory School, with his parents having to purchase his third school uniform in as many years.

One of Allan's pleasurable activities was to explore the harbour, where he was fascinated with the tug boat named John Dock and the ships that came from all over the world to offload their cargo. It was at this time that he developed his skills as a raconteur.[36] One day he returned excitedly from the docks and announced to his family that he had just enjoyed the most amazing experience. He told them how as he approached one of the large ships the captain had come down the gang plank to the quay, invited him to climb on board and given him a guided tour of the vessel. His parents and older sister were suitably impressed but, on further enquiry about what he had seen, Allan could no longer sustain the charade, burst into laughter and revealed that he had been pulling their legs! Another amusing incident occurred in their flat when Amelia discovered a puddle of urine next to the toilet bowl.[37] Convinced that the culprit must have been young Allan, she admonished him but he vigorously protested his innocence. On further investigation they discovered that the guilty party was Jock who, cooped up in the flat with nowhere to relieve himself, had chosen to urinate in the bathroom next to the toilet!

The Cormack family had been in Port Elizabeth for less than two years when George, who smoked a pipe though not cigarettes, contracted emphysema and died on 7 May 1936 (Figure 1.2). He was just 51 years old. Although he had worked for the South African Post Office for more than thirty years, there was not a substantial pension to support Amelia and her children. Bill was by now independent and had moved to Johannesburg to the University of Witwatersrand, Amy had just matriculated at Collegiate School for Girls while Allan's high school education still lay ahead. At the end of 1936 the three As – Amelia, Amy and Allan – together with Jock

Figure 1.9   George, Allan, Jock and Amy on the road from Port Elizabeth to Uitenhage, 1934.

returned to Cape Town, to the suburb of Mowbray beneath Devil's Peak, the Eastern 'bookend' to Table Mountain. Amelia was determined to give her son every opportunity to succeed in life and she enrolled him at Rondebosch Boys' High School, just two miles south of Mowbray and a single stop on the railway line. Given the family's financial position, Amy was unable to continue her studies at university and took a position with the Post Office Savings Bank.

In spite of their straitened circumstances, Amelia Cormack and her two younger children, Amy and Allan, set out on a European odyssey. For just £39 each they had secured a round-trip passage on the German passenger ship the *Windhuk*.[38] Setting sail from Cape Town harbour in March of 1938, they were treated to comfortable accommodation and wholesome German food, with Frankfurter sausages among the family's preferred cuisine. One of Allan's most enjoyable pastimes on board ship was playing chess. His ebullient self-confidence seemed to have no bounds and he would take on all comers. Towards the end of the trip he was presented with a chess set at one of the parties for the young passengers.

Once the Windhuk reached Europe, their first port of call was Amsterdam where the Cormack family disembarked for a short visit to the city. Their tour included the Rijksmuseum, the famous art gallery housing the great Dutch painters such as Rembrandt and van Gogh, which particularly impressed young Allan.[39] They also visited the Colonial Institute and the Royal Palace before boarding the ship which brought them to Hamburg, their first time on German soil.

From Hamburg they caught the train for Berlin where they stayed in a comfortable but quiet hotel. In fact, it was almost empty and gave them a rather eerie feeling when they first entered the hotel lobby. While his mother and older sister unpacked their bags in the room, Allan explored the foyer. After a short while he came racing back to the room, excitedly sharing the news with them that Adolf Hitler had stayed in their room some years before. This of course was a great talking point when the family later returned to Cape Town. The Cormacks could see quite plainly that the country was preparing for war. Soldiers were everywhere in evidence and there was furious activity all through the nights. The family noticed workmen digging in the streets, which they presumed was preparation for bomb shelters. When travelling around Berlin by bus they spoke to English-speaking locals. One particularly friendly woman in her mid-30s described how she and her family had suffered during the Great War of 1914–18 and said she was praying there would not be another war.[40] In less than 18 months her country would invade Poland and precipitate a conflict with far-reaching consequences for the world.[41]

Despite the dark omens, the Cormacks found Berlin to be a beautiful city. They toured the Reichstag (Figure 1.10), the Bismarck Memorial, the Brandenburg Gate and the National Art Gallery, which Allan described as a "centre of Nazi art".[42] Other sites included Unter Den Linden, the State Theatre, the Palace, the Lustgarten – identified by Allan as the venue for many Nazi speeches – and the Broadcast Tower and Exhibition Grounds. Allan was intrigued by the fact that half-way up the tower was a café with capacity for 100 people. Also on their itinerary was the Olympic Stadium, home to the Berlin Olympics in the summer of 1936. At the entrance were two tall towers between which the official Olympic symbol – five interlocking rings – was suspended on steel guide wires (Figure 1.10). The irony of the rings and all that they represented, in close juxtaposition with the Nazi swastika adorning one of the towers, was not lost on the young Allan Cormack.

Other highlights for the family in Germany were a boat ride on the Wannsee, a tour of the Sans Souci Palace, where they were accompanied by a group of French tourists (referred to somewhat mischievously by 14-year-old Allan as "Froggies") and a visit to Hagenbeck's Zoological Gardens. There Allan had his photograph taken, seated on a bench with two young

Figure 1.10   Berlin in 1938: The Reichstag (left) and Olympic Stadium (right).

lion cubs sprawled across his lap. This image of the African-born teenager with the acknowledged king of beasts from the African prairie appeared as the centrepiece for a story about Allan in a Japanese children's magazine over fifty years later (see Appendix E).[43]

Amelia Cormack and her two younger children next travelled to the United Kingdom, disembarking at Southampton. From there they headed north by train, stopping off in Edinburgh to visit some of Mrs Cormack's colleagues from her days as a student teacher at Moray House College. Allan and Amy were relieved that one or two of their mother's friends had passed on, meaning one less visit for them![44] Their final destination was the county of Caithness (Figure 1.1). There they again stayed at 54 Argyle Square in Wick, the home of Thomasina MacLeod, Amelia's mother, together with her bachelor son John and spinster daughter Betty. Amelia's other brother Allan lived in Cullen, in the county of Banffshire, and was estranged from John, Betty and their mother Thomasina for reasons that were later described by Allan Cormack as "mysterious".[45] South Road, Wick, was the home of the Dunnetts, whose two sons played with Allan. Uncle John took Allan and Ian Dunnett to play golf at Reiss Golf Club. The course was apparently infested with rabbits and according to his sister, Allan had more fun trying to catch them than playing golf!

Although this was a time for him to enjoy his Caithness roots with his mother Amelia (Figure 1.11), most of the knowledge that Allan absorbed about Scotland was a highly romantic view of its history as told by his Uncle John, who was a keen Scottish Nationalist. This view was reinforced by the many ruined castles, Pict houses, chambered tombs and standing stones visible in the treeless Caithness landscape. Clearly this pointed to a history stretching back several thousand years. During their summer in Wick, Allan and Amy were able to explore much of Scotland, an experience that no doubt led to their life-long affinity with the country.

Figure 1.11   Allan and his mother Amelia at Nairn in Scotland, 1938.

A beautifully illustrated colour map of Scotland, hand drawn by Allan Cormack, adorns the frontispiece of his 1938 album and shows the route taken by the family as they explored the countryside. Their journey began with a ferry ride from Wick to the Orkney Islands, a stormy passage across the Pentland Firth that left Allan seasick though, judging by the photograph, in good spirits. There they explored Kirkwall, with its impressive cathedral and tall steeple, the Earl's Palace and Stromness, a picturesque village with connecting waterways known as a 'Northern Venice'. Just north of Stromness, the second largest of Orkneys' towns, they visited the prehistoric settlement at Skara Brae and the Standing Stones of Stenness.

Back on the mainland Uncle John took them to Thurso where they explored the castle, home of Sir Archibald Sinclair, a distant relative of the Cormack family. The family then headed south and west, passing through the Western Highlands at twilight, said by Allan to be "one of the most beautiful scenes I have ever seen",[46] to the port of Ullapool where they clambered aboard a schooner. Next stop was Inverness, capital of the Highlands, and then on to Drumnadrochit with Loch Ness nearby. Alas, the monster did not make an appearance for the expectant teenager!

Travelling further south the family reached Fort William, scenically located on Loch Linnhe with Britain's tallest mountain, Ben Nevis, looming in the background. Uncle John regaled the Cormack siblings with the legend of Bonnie Prince Charlie who had landed at Loch Shiel near Fort William in 1745 and attacked the English. They also visited a beautiful waterfall in the Dark Mile, near Achnacarry, that served as Prince Charlie's escape route after the disaster of Culloden in 1748. Another historical site that they visited was Glencoe, the Glen of Weeping and the scene of the massacre of the MacDonald clan by the Campbells in 1650. Glencoe was significant for Allan because it was the birthplace of his great-grandfather Murray, his father George's maternal grandfather (Figure 1.2).

They drove down the western edge of Loch Lomond and came to their final destination, Glasgow, which was hosting the Empire Exhibition of 1938. Bellahouston Park had been transformed into an international exposition that incorporated purpose-built pavilions from all the countries of the British Empire. Although the architectural theme tended towards Modernism, the South African pavilion was designed in the colonial Cape Dutch style, with stark white walls, solid wooden doors and the characteristic stepped gable. Its exhibits included diamonds, ostrich feathers and indigenous handicrafts. For Allan the Exhibition was a marvellous educational experience with its impressive displays such as the Palace of Engineering, and it also connected the land of his birth with the land of his forbears. It was a fitting end to a fascinating exploration of his Scottish heritage.

The return trip from Southampton was aboard the S.S. Pretoria, the two week crossing allowing the family to reflect on their European odyssey and the re-acquaintance with their Scottish roots. Back in Cape Town, Allan returned to his studies at Rondebosch Boys' High School. However, since he had missed over three months during the middle of the school year (which runs from late January to early December in South Africa), he was not promoted to the next class at the end of 1938.[47] This turned out to be fortuitous because the following year he secured an academic scholarship that relieved the financial pressure on his mother. In addition, he was able to devote himself to his studies in his last three years of high school (1939–41), and thus prepare himself for a university education.

Chapter 2

On the Slopes of Table Mountain

> Below the cable station is a ledge called Fountain Ledge and
> this last 400 feet of Table Mountain must contain some of the
> finest rock climbing in the world.
>
> Allan Cormack[1]

Fourteen-year-old Allan Cormack was privileged to have the love and support of his family, although it would not have been easy finding his way in the world without the presence of his father. He was fortunate to have been enrolled at Rondebosch Boys' High School (RBHS), a government school located near the slopes of Table Mountain (Figure 2.1). The school had the benefit of being served by a generation of fine teachers who made education their life's work. Many years after his return from Scotland to complete his schooling at RBHS, Allan recalled with affection the influence that these men had had on him:

> Pictures of the staff of 1944 brought back many happy memories. Almost all of those staff members tried to teach me one thing or another at one time or another and I am grateful to them for their efforts. I can think of very little that I was taught which has not served me well over the years.[2]

Prominent among these dedicated teachers was Walter George Amos Mears who was headmaster of RBHS from 1930 to 1951. Born in 1890, Wally Mears was the son of a Methodist missionary, the Reverend William Mears, and was brought up near Umtata in the Transkei.[3] He grew up speaking Xhosa, the language of the local black population, having been

23

taught by his native-speaking playmates.[4] When he was old enough, Mears was sent to Kingswood College, a Methodist school in Grahamstown, and then to the recently established Rhodes University College in the same city. His education was completed at Emmanuel College, Cambridge, where he earned an honours degree in modern history and later a master's in education with a focus on psychology.[5]

In addition to his responsibilities as headmaster of RBHS, Wally Mears taught history to the top sections of the two highest classes in the school. The curriculum was mandated by the province's educational establishment which even then, in the late 1930s, was dominated by the descendants of Cape Afrikaners. So, much of what he was obliged to teach was 18th and 19th century South African history, which Allan characterised as "very dull stuff", endless Frontier Wars and constitutional affairs.[6] Given his upbringing in the Transkei and his Cambridge education, Mears' interpretation of history differed sharply from that of the authorities and so his teaching deviated from the standard curriculum. The result was that students whose examination scripts might have received a first class pass from Mears, when marked by the official examiners, received a third class pass and worse. Since performance in the matriculation examination determined entry into South African universities, Mears withdrew from teaching history so as not to prejudice the prospects of his pupils.

Although many in the school community attached great importance to winning matches on the playing fields, Mears disputed the emphasis on winning, espousing an educational philosophy that emphasised the spiritual benefits of participation in sport.[7] He strove to produce well-rounded individuals of high integrity and strong moral character. Although it was a government school, there was a strong Christian influence at RBHS. Mears and his wife identified themselves very closely with the rights of black South Africans living in the Western Cape. Although black children were denied the opportunity of attending RBHS, Wally and Nan Mears were active in the African Scholars Fund and played a leading role in the establishment of Langa High School in a black township on the outskirts of Cape Town. This commitment to the wellbeing of the underprivileged left a lasting impression on Allan and his fellow pupils.

In the classroom, among Allan's favourite subjects were the natural sciences – mathematics and physics – taught by Derham 'Jooks' Jackman and AA Jayes. Forty years after his RBHS days, and just a few months before receiving the telegram from the Nobel Foundation in Stockholm, Allan wrote a letter to his *alma mater* that appeared in the school's newsletter:

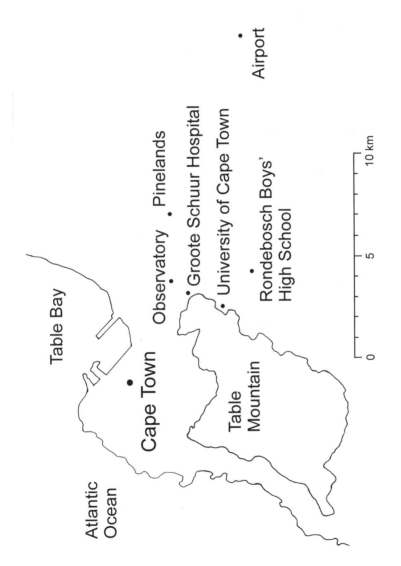

Figure 2.1   On the slopes of Table Mountain.

> As a matter of fact I recently published a short analysis of an
> aspect of CAT-scanning systems in which the only mathematics
> used was an elementary theorem on Euclid which I first learnt
> from either Mr Jayes or Mr Jackman![8]

Jooks Jackman had entered service at RBHS just before the First World
War, appointed directly from Great Britain. Although his career was drawing to a close in the early 1940s, he nevertheless left an indelible mark on
the young Cormack. Jackman was described as a "practical and forceful
teacher of science" and was known for his commitment to sport, particularly
swimming and track and field athletics. With his broad Yorkshire accent,
he was quick to "dooble" any homework assignment not completed on time
or, when circumstances dictated, he would instruct a pupil to "Bend doon,
boy!"[9]

AA Jayes, who in 1936 joined the staff of RBHS from the South African
College School, was renowned as a teacher of mathematics and chemistry
but it was in pure physics that he placed the school on the educational
map.[10] He successfully built up the school's laboratory apparatus so that, in
time, the RBHS facilities were admired by science teachers from the Western Cape and further afield. His commitment to and passion for his subject
were widely respected while his obvious skill and panache as a demonstrator
of physical phenomena inspired a generation of aspiring scientists (Figure
2.2). Some of his pupils, including Allan, would go on to enjoy successful
careers in academia. Later in his career at RBHS Jayes was to be appointed
vice-principal, where his commanding presence brought both dignity and
high standards of behaviour among the boys.

In mid-1939, at the age of 15, Allan was appointed one of two Junior
Librarians. While he was browsing in the school library he came across the
popular writings of two of Britain's foremost scientists of the 20th century,
Sir Arthur Eddington and James Jeans, both of whom were knighted for
their services to science. Their books opened Allan's eyes to the wonders
of the physical world.[11]

Educated at Trinity College, Cambridge, where he was the Senior Wran-
gler in the Mathematical Tripos, Eddington subsequently concentrated on
astronomy at the Royal Observatory in London. There he introduced a
method for analysing photographic observations of the stars to determine
an accurate value for the solar parallax.[12] Following his appointment as the
Plumian Professor of Astronomy at Cambridge in 1913, he became interested in the theory of general relativity when he received papers by Albert
Einstein from the Royal Astronomical Society. In 1919 Eddington led an

Figure 2.2   AA Jayes was an inspiring science teacher at RBHS

expedition to Principe Island off the coast of West Africa where he made observations of stars close to the sun during a total eclipse, thus verifying the bending of light as predicted by relativity theory. He would go on to publish the book *Mathematical Theory of Relativity* in 1923, which was described by Einstein as "the finest presentation of the subject in any language".[13] Eddington grappled with the challenge of uniting the theories of quantum mechanics and general relativity while it was his book *The Nature of the Physical World* that caught the attention of the young teenager in Cape Town.

James Jeans was five years older than Eddington and also studied mathematics at Trinity College, Cambridge. He was a natural writer and in the early part of the 20th century he published on a variety of topics in applied mathematics, physics and astronomy. There was a long running scientific rivalry between Jeans and Eddington over the mechanism by which energy was created in the stars. Much to the chagrin of Jeans, Eddington's theory, the gradual annihilation of matter, would prove to be correct.[14] Allan read *The Mysterious Universe*, published by Jeans in 1930, and was struck by the statement:

> We have already considered with disfavour the possibility of the

universe having been planned by a biologist or an engineer; from
the intrinsic evidence of the creation, the great Architect of the
Universe now begins to appear as a pure mathematician.[15]

Eddington and Jeans thus spurred Cormack's interest in mathematics
and physics, from the nucleus to the cosmos. He was enthused by the new
theories of relativity and quantum mechanics and certain that his future
career would be in science.

In Europe, Hitler and his Nazi Party vowed to avenge Germany's defeat
in the Great War and the subsequent humiliation of Versailles.[16] As the
move towards another conflict gathered momentum in the northern hemisphere, Afrikaner nationalism suffered a setback in South Africa.[17] General
Herzog wanted the country to remain neutral but Jan Smuts argued that
South Africa should side with Britain. Smuts won a bitter debate in Parliament and became Prime Minister again in July 1939.

The RBHS magazine of November 1939 recorded the outbreak of hostilities and an editorial spoke about the need for Great Britain and her free
Dominions to take up the sword in defence of democratic ideals. The editor, Mr AJA 'Cocky' Wilson, displayed a tendency towards the dramatic,
reminiscent of Winston Churchill (who would replace Neville Chamberlain
as Prime Minister of Britain six months later in May 1940) at the height
of his oratory powers:

> We cannot doubt the ultimate issue. Truth, justice, freedom,
> tolerance and unselfishness, these, though stubbornly assailed,
> shall never perish from the earth. Though the struggle may be
> long and the sacrifice great, we shall prevail.[18]

There was a separate section in the magazine, entitled 'The School at
War', in which the whereabouts and activities of Old Boys who had signed
up for the Allied war effort were reported. Inevitably, of course, this section would record the names of those who had made the ultimate sacrifice.
The RBHS Cadet Corps was resuscitated, with the masters serving as officers, and the senior boys as warrant officers and non-commissioned officers.
The South African Defence Department provided instructors, built a rifle
range and conducted training camps for the cadets during the school holi-
days. One of these camps was held at Pollsmoor (later the site of a prison
which subsequently gained notoriety as the home of ANC leader Nelson
Mandela), where Allan received the training that led to his appointment as
Company Quartermaster Sergeant.[19] The commanding officer of B Company was Captain Wilson, the selfsame Cocky Wilson and editor of the

RBHS magazine, who wrote these melodramatic verses:

Boys of the School, in every land
Where Freedom's flag is still her pride,
We stretch to you a comrade's hand
And stand with you, whate'er betide.

Your courage is our courage, too,
What you have dared we too shall dare.
When comes the time to stand with you
Boys of the School, we shall be there![20]

The Debating Society played a prominent role in the lives of many boys at RBHS. Aside from the obvious benefit of improving their skills and confidence in public speaking, the debates had a positive educational spin-off. The topics chosen were varied, from 'Science is a truer benefit to man than religion' and 'Man is born for war and not for peace', to 'South Africa's native policy does not justify the inclusion of the Protectorates'.[21] (The Protectorates referred to the countries of Basutoland, Bechuanaland and Swaziland – see Figure 1.6 – which were all under British protection.) The headmaster Wally Mears served as President of the Society. His influence on the selection of topics was evident, while he and his wife Nan provided refreshments after the meetings. Among the highlights of the season were the inter-school debates against Bishops and Herschel, private schools for boys and girls respectively. The topic for the debate against their sister government school, Rustenburg Girls' High, 'That there should be equal pay for equal work for men and women', drew a crowd of almost one hundred.[22] Many outstanding orators and hecklers were present and, after a thorough discussion, the motion was carried by just three votes, 49 – 46.

Another popular format was the 'hat debate'. Topics, of which the boys had no prior knowledge, were written on slips of paper and placed in a hat. Each member of the Society would then be invited in turn to draw a slip from the hat and, after a brief glance at the topic, deliver an impromptu speech lasting two to three minutes. In his final year at school Allan served as the elected secretary of the Debating Society (Figure 2.3) and his report in the RBHS Magazine of December 1941 described how the new members, with little prior experience of public speaking, had found their feet and revealed some latent talent.[23] In Allan's own case, his participation undoubtedly honed his skills as a raconteur and stood him in good stead for his later career as a university lecturer.

Figure 2.3   Allan Cormack (centre) as secretary of the Debating Society with the two masters in charge: Wally Mears (far right) and Jack Kent (far left), 1941.

Cocky Wilson had joined the RBHS staff in 1913 and his 33 years of service to the school were interrupted only by three years of active duty in East Africa during the First World War. In stature he was short and stout, while professionally he was recognised as an excellent English teacher and a talented poet and writer.[24] As a graduate of the University of London, Wilson sometimes became agitated by the clipped accents of his South African pupils. He would occasionally parody their recital of the Lord's Prayer during morning assembly by later writing on the black board what he had heard, "Ow far chart in ebben". Wilson's passion was the theatre and his indefatigable leadership during the 1920s and 1930s led to the establishment of the school's reputation for dramatic productions. In all he produced 25 full-length plays during his time at the school, including Shakespeare (often dictated by the matriculation set work for that year) and more light-hearted works such as the vaudeville show *For Victory*.

A major challenge in mounting a production was to overcome the limitations of a small cramped stage in the school's hall. One of Allan's fellow pupils with an engineering bent devised a primitive dimming system for the lights.[25] In this contraption, an electric current was passed through a drainpipe that was filled with water. By raising or lowering a plunger attached to a length of string, the electrical resistance was regulated and the brightness of the light shining onto the stage was controlled. The system worked well but one evening, during the production of *The Admirable Crichton*, disaster struck. The string broke, the plunger fell precipitously and a scene which required a gently fading twilight was suddenly replaced by dazzling sunshine! The school called in the municipality's Electrical Department to repair the damaged wiring and the technicians were literally shocked by the poor state of the electrical system!

In the winter of 1941 RBHS joined forces with its sister school, Rusten-burg Girls' High School, to mount a production of the play *Dear Brutus*, written by JM Barrie in 1917. Barrie, who also wrote *Peter Pan*, had been inspired by Shakespeare's *A Midsummer Night's Dream* when writing the play. In Act I, a group of eight strangers are assembled at a country house on a midsummer's eve, their host a puckish gentleman called Lob. Each of the guests reveals the conflict in his or her life. Lob invites them all to join him for a walk in a mysterious forest where they will get "what nearly everybody is looking for: a second chance". Act II takes place entirely in the woods and Barrie weaves his web, creating a tale of 'might-have-beens' with serendipity and irony. The players return to the mansion and Lob's drawing room in Act III, to the lives they left behind when they ventured into the woods. Sadder, yet wiser, they come to realise what has happened to them. As each character grapples with the challenge of making a fresh start, Barrie's optimistic ending reminds the audience that in some way they are all lost children.

This was a difficult play for the youthful players from the two schools, most of whom had never performed on the stage before. Cocky Wilson produced the play and was impressed with the "excellent blend of natural talent and acquired stage-craft".[26] Mrs Maisie Wray created the attractive Edwardian costumes, Mr G Cappon was responsible for the settings which added the right atmosphere to the drawing-room and forest scenes, while the RBHS schoolboys were in charge of the lighting effects. Fortunately for the production of *Dear Brutus*, there were no problems with the dimming system.

Allan played the part of William Dearth, an artist whose marriage is failing and, as his painting skills diminish, finds refuge in alcohol. His sympathetic portrayal of Dearth was recognised by the *Cape Argus*, which judged his scene in the woods with his 'might-have-been' daughter Mar-garet, played by Maureen Strauss, to have been the best in the play.[27] Allan's part also required him to smoke his first cigarette and, as his older sister Amy lamented, this became a habit that "continued far too long thereafter!"[28]

The opening night for *Dear Brutus* was Friday 13 June 1941 and, over the following five days, a further four performances were mounted. The ticket sales were donated to two deserving war charities, the Bombed Cities of Britain and the Merchant Navy Fund, and £115 was raised for each. The music was provided by the RBHS orchestra and an atmospheric piece which served as the prelude to Act II was specially composed for the production by Mr AR 'Foxy' Howse, an inspiring teacher of mathematics and a talented musician.[29] The combined efforts of the two schools resulted in a successful production of the play that could have been staged in no other way (Figure 2.4). For Allan, it provided another opportunity to develop his confidence on a public stage.

Figure 2.4   The cast of *Dear Brutus*, with Allan Cormack fourth from left, June 1941.

Aside from his involvement in the dramatic productions, Debating Society and the Cadet Corps, Allan's final year at RBHS was also concentrated on his preparations for university. At mid-year he and his family received the news that he had been awarded the Mason Scholarship, named in honour of a previous headmaster.[30] As Senior Librarian he was able to set aside precious time in the school's library, immersing himself in his books and preparing for the upcoming matriculation exams. In November 1941, a total of 56 boys wrote and passed the examination, seven of them with a First Class pass. Allan Cormack was among those seven.[31] He and his fellow matriculants could now look forward to a well-earned break over the upcoming Christmas period before embarking on the next phase of their lives.

Inspired by the writings of Eddington and Jeans, Allan briefly considered the possibility of pursuing a career in astronomy. He perceived his chances to be extremely slim, however, unless he was able to graduate with a very good degree in mathematics and physics, preferably from Cambridge

University.[32] He recognised that RBHS was a very small pond and whatever success he had gained at school was no guarantee of future success in the considerably larger ponds of South African universities. Although he held his own teachers in high esteem, the thought of teaching at a high school, if a career in astronomy failed to materialise, did not appeal to him at all. Astronomy, therefore, seemed to be out of the question. On 7 December 1941, the Japanese air force attacked without warning the American fleet at anchor in Pearl Harbor in Hawaii. World War II had become a global conflict.[33] While many young South Africans continued to volunteer for the Allied war effort, Allan was under the age recommended for enlistment and chose to continue with his education.

On the morning of 17 February 1942 Allan left the flat that he shared with his mother and sister – at 15 Brays Court, Station Road, Mowbray – and went to register for his first year of studies at the University of Cape Town (UCT). The main campus was located in the suburb of Rondebosch, less than two miles south of Mowbray on the eastern escarpment of the Table Mountain range (Figure 2.5). As he walked slowly up the slopes of Devil's Peak to register, he was uncertain what degree course he should follow. His father had been a telecommunications specialist in the engineering department of the Post Office while his older brother Bill had trained as an electrical engineer at UCT. He had noticed that there seemed to be no shortage of job openings in the field and so late in the morning Allan registered for a four-year degree in electrical engineering. The required courses were pure mathematics, applied mathematics, physics and engineering geometry.[34]

When Allan returned home later that afternoon and told his mother of his choice of study she expressed some surprise and smiled at him, in a kindly way. Amelia Cormack knew that her husband and elder son had been fascinated by machines and all manner of gadgetry, both mechanical and electrical, and that her younger son did not share this intense interest. Despite her misgivings, Allan embraced his studies with enthusiasm and at year's end he was rewarded with a first class pass in three of his courses, the only blemish being a second for pure mathematics.[35]

In May 1942, during Allan Cormack's first semester, there appeared in the local press editorials that were critical of UCT's students, suggesting that some were shirking their duty to South Africa by not enlisting in the war effort. Stung by this criticism, the university's own student publication, *Varsity*, reported that final year engineering students had sent a memorandum to the Director of Recruiting, offering to complete their training via an

Figure 2.5   University of Cape Town on the slopes of Table Mountain, 1945.

accelerated curriculum so that they could be ready for service by mid-year.
The telegraphed reply from the military authorities stated that no urgent
necessity for this course of action existed.[36] UCT's Principal and Vice-
Chancellor, Dr Arthur W Falconer, reported that students in the third or
later years of their curriculum were encouraged by the military authorities
to complete their degrees before enlistment.[37] He also stated that individuals under the age of 19 – whether students or not – were discouraged from
joining the military. Allan had celebrated his 18th birthday just a week before registration, and he was therefore not obliged to enlist. Although 225
male students at UCT were eligible for call-up, Falconer cautioned against
the dangers of a policy that differentiated among its students.

Within UCT there had been intense debates during the 1930s about
the nature of a university and what should be taught. A few students
were themselves concerned that the university was becoming a 'glorified
technical college', with the Professor of Electrical Engineering, Brian L
Goodlet, paradoxically noting that:

> The true university atmosphere and spirit of disinterested in-
> quiry is menaced by the large number of students who nowadays

come to the university simply to be instructed in some useful art whereby they can earn a living.[38]

Recruited from the United Kingdom to UCT in 1937, the 34-year-old Goodlet led a renaissance in the Department of Electrical Engineering, updating the syllabus so that major developments in the field of light current, including electronics, were introduced. Nicknamed 'Prote' after the well-known American engineer Charles Proteus Steinmetz (who was a pioneer in the development of alternating current), the Cambridge-trained Goodlet brought an intellectual vibrancy to bear on his department.[39] Shortly after the outbreak of World War II he took a leave of absence from UCT, first at Birmingham and then with the South African and Royal Navies to contribute to the war effort.

The Germans were laying magnetic mines on the seas of the southern hemisphere and Commander Goodlet recruited students from UCT's electrical engineering and physics departments to assist with the 'degaussing' of ships for all latitudes and longitudes.[40] Among these students was Godfrey Stafford, an alumnus of RBHS, who would later become one of Allan Cormack's special friends. Goodlet recognised that the solution to novel wartime problems, such as magnetic mines and radar, required engineers who were trained in research and had a strong foundation in the physical sciences. By the time Allan signed up for electrical engineering in February 1942, Goodlet had introduced a second option, the B-curriculum, in which much of the traditional engineering material was replaced by courses with an emphasis on physics and mathematics. Allan soon switched to the new curriculum and quickly realised that he not only enjoyed the courses, but excelled in them as well.[41]

In applied mathematics he studied statics, rigid body dynamics, and spherical trigonometry and was then introduced to relativity theory in his third year.[42] There was also a section on astronomy, covering planetary motion, parallax, precession and aberration, which Allan particularly enjoyed. The curriculum in pure mathematics in his first year consisted of elementary differential and integral calculus, and co-ordinate geometry in two and three dimensions. By the third year, he and his colleagues were grappling with Fourier series, theory of numbers, calculus of variations, as well as ordinary and partial differential equations. In physics, Allan studied the kinetic theory of gases, thermodynamics, physical optics, electromagnetic theory, alternating currents and elementary atomic theory. There was also a practical course which consisted of experiments on these subjects, together with training in glass- and metal-working and laboratory techniques.

In 1943, Allan's second year at UCT, he began to play an active role in extracurricular activities, joining the Engineering and Scientific Society and being elected as the second year representative.[43] The President of the society was Professor Frederick Walker, head of the Department of Mineralogy and Geology. In his presidential address to the society that year, Walker stressed the need for cross-fertilisation between the different academic disciplines. He spoke on 'Geology and Engineering' and showed how millions of pounds could have been saved if the engineers had consulted geologists before drilling in vain for water or petroleum, or constructing aerodromes that were rendered unusable after a thunderstorm.[44] Walker called for an end to the days of intellectual snobbery, a message that clearly had an impact on Allan who, two years later, would choose a master's thesis that was located at the interface between physics and geology.

Allan's extracurricular activities probably took their toll on his academic performance for at the end of his second year his grades had slipped slightly, and he earned second class passes in all subjects except Physics II, for which he received a first. Because he had been following the B-curriculum in electrical engineering, Allan was able to switch his major to physics in March 1944, the beginning of his third year at UCT. Interestingly, he was permitted to take a first year chemistry course – a standard requirement for the pure BSc degree – on condition that he completed special work in applied mathematics to the satisfaction of the head of the Department of Applied Maths.[45] Not only did he achieve a first in Chemistry I that year, but he was also awarded his degree with a distinction in physics.

Reginald William James had been appointed Professor of Physics at UCT in 1937, replacing the brilliant Basil Schonland, whose investigations into lightning had established an international reputation that led to him taking up a prestigious appointment at the University of the Witwatersrand in Johannesburg. James had come to Cape Town partly as a result of the machinations in British academic politics. He held the title of Reader at the University of Manchester, having established himself both as a fine teacher and a successful researcher, during which time he elucidated the basic interactions of X-rays with crystals. He was therefore in line for a chair in physics at one of the British universities but, as they became vacant, Ernest Rutherford, the Cavendish Professor at Cambridge University, was filling them with 'his boys'.[46] James therefore was obliged to do what others had done before him – including Rutherford himself and WH Bragg – and accept a chair in one of 'the Colonies', at Cape Town.

Not all of Rutherford's 'boys' secured the expected appointment, however. James enjoyed telling the story about the chair in physics at the University of Aberdeen which had become vacant in the mid-1930s.[47] James was an applicant, as was PMS Blackett, Rutherford's preferred candidate who would later win a Nobel Prize, and a third short listed applicant. Blackett was the overwhelming favourite to be offered the chair, but inexplicably and to everyone's surprise it went to the third candidate. Shortly afterwards James was at a scientific meeting in London when he bumped into the ebullient Ernest Rutherford. Rutherford was of course aware that James had been a candidate for the post in Aberdeen and, before the meeting started, greeted him in a loud voice that was clearly audible to all those present: "Hullo, James!" he said, "Extraordinary appointment at Aberdeen, wasn't it? Extraordinary!" Such affirmation by the great physicist provided James with a certain degree of satisfaction which he shared with his young lecturing staff in the tea room at UCT over a decade later.

James established a crystallography group at UCT, since the equipment needed was relatively inexpensive and the technology leant itself to the solution of locally relevant problems in geology and mining.[48] Despite carrying a heavy teaching load – he taught the large Physics I class single-handedly – James managed to inspire his students with the excitement of research. Even in the early 1940s he recognised the potential of crystallography to elucidate the structures of large biological molecules, telling his physics students that therein lay the future of biology. His effect on Allan was profound:

> James was a marvellous teacher, an excellent physicist, and a man of the highest intellectual and moral probity.[49]

Allan graduated with his BSc in Physics a week before Christmas on 18 December 1944 and, having been inspired by his mentor RW James, he decided to return to UCT the following year to pursue an MSc. Before taking that next step, however, Allan had a two-month summer vacation to enjoy. He approached Dr John Jackson, His Majesty's Astronomer at the Royal Observatory, Cape of Good Hope, and asked him for a vacation job. Allan was assigned to Jackson's first assistant, Richard Hugh Stoy, who was the Deputy Director of the Cape Observatory, conveniently located less than a mile north of the Cormacks' flat in Station Road, Mowbray. Finally, he would get an opportunity to indulge himself, to experience the excitement of conducting research in the field of astronomy, and to immerse himself in the world of his boyhood heroes, Eddington and Jeans.

For many years the astronomers at the Cape Observatory had been measuring the stellar parallaxes – distances calculated by triangulation in which the diameter of the earth's orbit served as a baseline – and had built up a significant archive of photographic plates on which were recorded images of many hundreds of stars. Following the lead of the astronomers at the Leander McCormick Observatory, run by the University of Virginia just outside Charlottesville, Virginia in the USA, Stoy and Allan decided to use the photographic plates for accurate measurement of the brightness of the stars recorded on them.[50] They were especially interested in the stars for which parallaxes had been measured. The archive of photographic plates gathered at the Cape Observatory over a period of thirty years thus provided a rich mine of raw material awaiting exploitation.

Stellar parallaxes were of interest to astronomers in the mid-1940s because they provided basic data essential to the study of stars. The parallaxes are still important today although the ground-based measurements made by Allan during his summer vacation over sixty years ago have been replaced by space-based observations such as those of the Hipparcos satellite. To test the quality of their results, Stoy and Allan selected two bright stars – Sirius and Canopus – for which there existed very carefully gathered measurements. There was excellent agreement with the published literature and, encouraged by the findings, Jackson communicated their manuscript to the Royal Astronomical Society in London. Later that year Allan experienced the thrill of seeing his name in print in his first academic publication, which appeared in the *Monthly Notices of the Royal Astronomical Society*.[51] Aside from Stoy and Jackson, Allan also interacted with Harold Knox-Shaw and Roderick Redman of the Radcliffe Observatory in Pretoria, who provided visual magnitudes for stars in the Cape Parallax and feedback on the draft manuscript. He discovered that all four men were Cambridge graduates, confirming his belief that this particular university must be the key to a successful career in astronomy.

The next year Allan joined the Natural Sciences Society and was elected to serve on the committee. The society's President was Professor James who, on Wednesday 23 May 1945, delivered his presidential address on 'Seeing Atoms and Molecules'.[52] In attendance were a hundred members and visitors who were treated to an erudite lecture on a difficult topic as James showed pictures of a molecule that had been drawn from photographs taken in the physics department. Just three months later he would again demonstrate his ability to explain complex physical phenomena to a general audience.

In Europe the Allies had begun their final assault on Germany in the spring of 1945. With the Russians advancing on Berlin from the east, the British swept into northern Germany via the Netherlands, while the French and American forces moved rapidly into central Germany from the south-west. On 30 April, deep underground in his Berlin bunker, Hitler committed suicide. Within a week the Germans had surrendered unconditionally and the war in Europe ended. The world's attention now turned to the conflict in south-east Asia and the Pacific.

Fearing that a full-scale assault on the Japanese mainland would cost many Americans their lives, President Harry Truman – the successor to Roosevelt who had died in April – made the fateful decision to deploy an atomic bomb. Japan was issued with an ultimatum: surrender unconditionally or face the prospect of destruction by a weapon whose power the world had not yet seen. The Japanese ignored the warning and chose to continue fighting. On 6 August an American B-29 bomber, the Enola Gay, dropped the first atomic bomb ever used in warfare on the city of Hiroshima. Almost 100,000 people, the vast majority of them civilians, were killed by the explosion which also destroyed a five-mile-square swathe of the city.[53] Three days later the second bomb was dropped on Nagasaki, resulting in further widespread death and destruction, and within a week the Japanese capitulated and agreed to halt the war.

At the University of Cape Town a symposium on the atomic bomb was arranged and attended by Allan and his friends. The presentations were reported in the student publication, *Varsity*. RW James explained how he thought the bomb might have worked. He summarised how the energy was derived from nuclear fission, or the splitting of the atomic nucleus of uranium into two parts. He described how, once sufficient quantities of the unstable isotope uranium 235 had been prepared, it was possible to propagate the action through the bulk of the material, each fission generating additional neutrons that carried the reaction to new atoms. James concluded that the method by which the action was initiated, and how the speeds of the neutrons were controlled, were obviously closely guarded secrets. George Stewart, a senior lecturer in the Department of Civil Engineering, spoke about the huge potential of nuclear power to contribute to the provision of electricity. The most telling commentary came from Marthinus Versveld, who would later become Professor of Ethics and Political Philosophy. He wrote, somewhat presciently:

The column of smoke that, while history endures, will hang

over Hiroshima, marks the culmination of an achievement of
our post-medieval Western society much more significant than
its scientific achievements: the extinction of the idea of man.[54]

For his MSc degree in physics in 1945 Allan Cormack was awarded the
Rodney Noble Prize, worth £80.[55] Since he was still living at the flat in
Mowbray with his mother and sister, these funds provided some much-
needed pocket money. The degree format included a series of advanced
courses – on topics such as thermodynamics, optics, electro-magnetic theory, radioactivity and quantum mechanics – followed by the performance of
experimental work of a research nature. Students were also encouraged to
take one or two courses in pure and applied mathematics and Allan availed
himself of this opportunity to broaden his expertise. For his master's thesis
work he chose to focus on crystallography, with Professor James as his supervisor. There were two PhD students who were studying complex organic
compounds but, because his time was limited, Allan elected to concentrate
on a series of compounds that were of interest to geologists.[56]

Professor DL Scholtz, head of the Department of Geology at Stellenbosch University, had written an extensive paper on the ore deposits, mostly
nickel and platinum group metals, that had been discovered at Insizwa in
the Mount Ayliff area, near the town of Kokstad in the Eastern Cape.[57]
These Parkerite minerals – nickel silicates and sulphides – were recognised
as having great potential and were being compared to the important ore
deposits at Sudbury in Ontario, Canada. Professor Scholtz provided Allan with a suite of 15 synthetic samples of the Parkerite series, ranging in
composition from bismuth to lead. Allan then used an X-ray camera to
characterise each of the powder samples, generating intensities and glanc-
ing angles for the lines of the crystal lattice. He published his findings in
the *Transactions of the Geological Society of South Africa*, concluding that
it would be necessary to re-investigate the chemical composition of the natural Parkerite from Sudbury in Canada before its likeness with the South
African Parkerite from Insizwa could be established.[58] Sadly, the body of
ore at Insizwa never lived up to its geological potential.

With World War II ending in 1945 there was an influx of ex-servicemen
to UCT at the beginning of the 1946 academic year. In the physics department the number of students grew to almost 700 and Professor James
was obliged to hire new staff so lectures could be duplicated and practicals
had to be offered on Saturday mornings. Having earned his MSc degree
with first class honours at the December graduation, Allan was offered a

temporary position as a junior lecturer in physics. Although postgraduate students constituted a small proportion of the student body, their achievements in the 1940s set the standard by which other academic departments, both at UCT and elsewhere, were judged. James himself, who a few years later would publish his definitive textbook, *The Optical Principles of the Diffraction of X-Rays*, remarked

> The outside reputation of a University does not rest on the excellence of its routine teaching, but on the reputation of its staff as investigators, and of the first rate students that it turns out.[59]

In the departmental staff room, Allan was influenced by two young lecturers who became lifelong friends. These were Godfrey Stafford, who had returned temporarily to UCT after wartime service with Commander Goodlet before heading off to Cambridge University, and John Juritz, a talented teacher in the mould of RW James, who would dedicate the following 40 years to the UCT physics department (Figure 2.6). A favourite pastime for the young staff during the midday break was a game of bridge which, on occasion, would extend well beyond the lunch hour.[60] Aside from coming to know some of the more established lecturers such as CBO Mohr, who returned to his native Australia in 1947 to take up a chair in theoretical physics at Melbourne, Allan was also exposed to the personal warmth of Professor James and his penchant for sharing a good yarn about his experiences in the United Kingdom. These anecdotes were told with obvious pleasure but with James's characteristic diffidence. Reginald James had established his reputation in physics at Manchester on the basis of his work on the zero-point energy in crystals. He presented a lecture on this work at a meeting of the Royal Society at which Ernest Rutherford was present. At the conclusion of the lecture he pointedly asked James if the experimental effect had been observed before or after the theoretical prediction. James had answered "Before, Sir" to which Rutherford responded enthusiastically, "That's the way it should be!"[61]

When Allan was an infant he had been diagnosed with an inguinal hernia that was of special concern to his mother. Although the problem resolved itself, Mrs Amelia Cormack was always worried that that the hernia might recur if he participated in vigorous exercise.[62] This was the primary reason why he did not participate in two of South Africa's national sports, cricket and rugby, when he was at high school. However, during his final years at RBHS he developed a keen interest in rock climbing, which he shared with

Figure 2.6   Cosmic ray experiment conducted by Godfrey Stafford as part of his MSc thesis; John Juritz and Allan Cormack assisted, 1946.

his best friend Walter Powrie.[63] The attributes demanded by this pursuit – upper body strength, balance, a head for heights, love of the outdoors, and the desire to seek new challenges – were well suited to Allan's temperament and outlook on life. When he entered his first year at the University of Cape Town in 1942 he joined the Mountain and Ski Club. Thus was born a passion for climbing.

Climbing in Cape Town in the 1940s required only the most basic of equipment: a 60-foot hemp rope and a pair of 'takkies' or tennis shoes.[64] The rope was initially cumbersome, stiff and difficult to coil but with frequent use became softer and much easier to handle. When the rope was new the breaking strain was as much as 2000 pounds, sufficient to carry the load of a dozen climbers; however, elasticity was virtually non-existent. Furthermore, the rope's material properties deteriorated rapidly with heavy use and there was always a fear that shock loading – such as a fall by even a small climber – could cause the rope to snap taut and break. This led to the old adage "the leader must not fall!"

Many of the favourite climbs were on the Western Table, the front face of Table Mountain. Few students had cars in those days which meant that Allan and his friends would catch the train from the southern suburbs to the main station in Cape Town and from there they would hop on the electrically-driven tram that left St George's Street and carried them up the hill to Kloof Nek. This was the gathering place for climbers in the early 1940s. Once they had set off, their next imperative was making morning tea, by all accounts a serious climbing ritual. It was practised at such shrines as Africa Ledge and Barrier Cave, before the really strenuous climbing began.

Within a few years of joining the UCT Mountain and Ski Club, Allan had established himself as a very accomplished climber, to the extent that he had begun to open a new route up the front face of Table Mountain. There were well established routes such as Sheerness Face, Union Route and Nerina Crag, but Allan was seeking a new challenge. With his undergraduate colleagues Walter Powrie and Chrystal Komlosi, plus Harry Biesheuvel, a lecturer in land surveying, he made his first determined attempt up the watercourse above Union Cave. They corkscrewed up a broken track, pinched out along a narrow band of rock from Platteklip Buttress, and then discovered a shoulder of assistance to launch themselves beyond a daunting overhang. Allan and Biesheuvel later described the final ascent in the *Journal of the Mountain Club of South Africa*:

> The ascent of this face is made in a shallow recess, the upper twenty feet being anything but easy. This section leads to a substantial ledge where at present a young pine provides an excellent belay. A crack can be followed to the top but it is strenuous, not too clean and the rock is doubtful in parts. The party traversed from the chockstone to the corner of the north wall of the chimney, some going straight up and the others traversing out on to the face and then up. Both alternatives are very sensational but most pleasing climbing. From the ledge reached some forty feet above the chockstone, a couple of short easy pitches lead to the top.[65]

It had taken many months for the route to be finally completed. That fact and the nature of the delays were sufficient reason for the route to be officially named *Procrastination Crag* (Figure 2.7). However, given his leadership and determination, Allan's friends always referred to it as *Cormack's Route*.[66]

One of Allan's regular climbing companions was Michael de Lisle, who was majoring in the classics. De Lisle was a few years older than Allan and, after studying at UCT for six months in 1940, had joined the Allied war effort.[67] He was captured at Tobruk in North Africa, along with many other South Africans, and held in an Italian prisoner of war camp. De Lisle managed to escape from his internment camp and was on the run for fifteen

Figure 2.7 *Procrastination Crag* (left) and Allan Cormack climbing on the cliff face (right), 1943.

months, behind enemy lines and reliant on the goodwill of ordinary Italian families – he had learned to speak Italian at UCT – before he managed to cross over to the Allied side. He returned to Cape Town at the end of 1944 and rejoined the University in January 1945. De Lisle brought the same intrepid confidence and *joie de vivre* that enabled him to survive the war years to his climbing exploits at UCT.

UCT owned a property in the Zuurberg mountains on which was located a small hut that was maintained by the Mountain and Ski Club. The club property was situated at the southern end of the Hex River Range, near the towns of Worcester and Rawsonville and less than 60 miles from Cape Town (Figure 1.6). Known as the Waaihoek Hut, this stone structure with an A-frame roof served as the base camp from which students could learn to ski on the nursery slopes or indulge their passion for climbing. On Monday 9 July 1945, during the mid-year University vacation, a group of students set out for a week's break to enjoy the recent fall of snow in the Hex River Mountains.

A party of eight caught the train to Breede River, divided up the food supplies on the station platform, and then hiked up to JG du Toit's farm

which lay beneath Waaihoek Peak. After spending the night on the farm in
the potato loft, they completed a six-hour slog up the slope to the hut, only
to discover the south window had been damaged in a storm and the whole
inside of the hut was covered in drifted snow. Weary from the climb, they
now faced the tedious task of clearing out the snow. Mauritz Boehmke
rewarded their efforts by serving up a hot curry stew, with apricot jam
serving as a substitute for chutney. It was not long after dinner that the
mountaineers, exhausted by the demands of the day, climbed into their
sleeping bags and fell fast sleep. An extract from Michael de Lisle's logbook, describing their climbing experience a few days later, reveals their
enjoyment of the breathtaking scenery:

> By Friday the snow was getting thin and heavily studded with
> rocks and lumps of ice, so Allan, Mauritz and I walked lazily
> over to Mount Superior in fine sunshine, climbed and unclimbed
> a tricky false summit – the 'Mother and Brat' – before finally
> traversing round a steep snow slope to the cairn where we left
> an envelope, with our names pricked on it with a knife-point,
> for lack of a pencil. The view was magnificent – towards Milner,
> the du Toit's Kloof fell steeply away from Superior, and on the
> Waaihoek side a sheer narrow gully running into the Waaihoek
> Kloof. Far to the North, over the left shoulder of the Gydoberg
> we saw a beautiful tabular mountain, sides black and vertical
> and a thick layer of snow on top, gleaming in the sunshine.[68]

For the first two days of their visit there was sufficient snow for them to
enjoy skiing in the vicinity of the hut. Then on Friday 13 July, one of the
women in their group, Mariana Gorm, decided she would take advantage
of the snow fall on the other side of the Waaihoek Peak. Unfortunately she
chose to go out alone. She climbed up to the beacon, strapped on her skis
and peered down at the slope in front of her. Mariana had never attempted
this skiing run before but, with little hesitation, she plunged down the
mountain. The slope grew steep very rapidly and she was soon hurtling
down at a speed that was difficult to control. Suddenly a chasm gaped
ahead of her and, realising the consequences, she deliberately angled her
skis and fell over. Her momentum carried her further down the mountain,
tumbling as she fell, but fortunately she was able to clutch onto a rock that
protruded through the snow and ice.

Bruised and in pain, Mariana began to crawl up to the summit. The ice
was hard and slippery and, without a crook, her progress was slow. After
five long hours she reached the top and, just as the sun was setting, the

exhausted but intrepid skier collapsed onto the floor of the Waaihoek Hut. Maurice Norton recorded Mariana Gorm's misadventure in a poem which captured perfectly the camaraderie that existed among the group of UCT students.[69]

While Mariana Gorm survived her ordeal on the Waaihoeksberg with a few cuts and bruises, not all encounters with the elements had such a happy ending. Just a month later, on Saturday 11 August 1945, tragedy struck the mountaineering community of the Western Cape. Under the leadership of 21-year-old UCT student Norman Fraser, a party of six set out in the face of a wild storm to climb up to the Waaihoek Hut. In the party was 13-year-old Roderick Thomson, a pupil at Rondebosch Boys' High School. He was on the verge of collapse when he returned alone in the dark, groping his way through the mist, and arrived at the farmhouse of Mr JG du Toit. Exhausted and soaked to the skin, he was incoherent and unable to explain what had happened. Mrs du Toit fed him a hot meal and put him to bed. The next day Thomson gave a first-hand account of his ordeal to the *Cape Times*.[70]

The party had left Cape Town in two cars at 7:30 on Saturday morning with the intention of climbing to the summit of Waaihoek and spending the night in the UCT hut. Fraser, who was a member of the UCT Mountain and Ski Club, was the only member of the party who had previously been to the hut. They departed from the base at 10 o'clock in poor weather – overcast, windy and a persistent drizzle – but were reluctant to abandon the trip, having driven all the way from Cape Town. John Bromback and young Thomson led the party and for the first three hours they made steady progress. Then their progress was slowed by the elements and three members of the party – Paul Kahn and his wife, and Mrs Queenie Becker – were struggling to keep up with the leaders. Fraser handed the key of the hut to Thomson so that he and Bromback could push ahead. By 4:00 pm they were in the snow belt and the drizzle had turned to sleet. Conditions by now were impossible and, estimating that they still had some distance to climb, they abandoned the ascent.

On the return trip John Bromback became delirious, his speech incoherent and he struggled to maintain his balance. Luckily the path was marked by beacons and so, in time, Thomson and Bromback linked up with Mr and Mrs Kahn. Kahn was also in a 'stumbling delirium' but fortunately his wife was still all right. She told Roderick Thomson that Norman Fraser and Queenie Becker had both turned back. She and Thomson managed to help the two men for a time but they were soon unable to make any progress at all, with Paul Kahn almost having to be carried. They decided to throw away their packs of food and search for a sheltered spot but were unable to find one. Mrs Kahn suggested to Roderick that he should return to the base at Vredehoek and call for help. On the morning of Monday 13 August, the following newspaper headline appeared:

Two Cape Town Mountain Climbers Were Frozen To Death During
The Weekend, And It Is Feared That Three Others
Including Two Women, Lost Their Lives At Waaihoek Mountain

A search party had gone out on the Sunday and returned to the du Toits' farm, Vredehoek, with the news that two bodies had been found high in the ramparts. When described to the boy, Thomson identified them as Paul Kahn and John Bromback. Their hands and legs were badly scratched, their clothes torn and Kahn was clutching an empty box of matches. The search party could find no trace of the other three climbers.

Chrystal Komlosy heard the news on the radio on Sunday evening and contacted her fellow UCT club members. Among those assembled for the second rescue party were Harry Biesheuvel, Oscar Keen, Mauritz Boehmke, Michael de Lisle and Allan Cormack. They reached the farm Vredehoek just after mid-night and started the search at daybreak on the Monday morning. Within a few hours they had reached the bodies of Kahn and Bromback. Allan was among the party of six that went ahead to the Waaihoek Hut.

As highlighted by Michael de Lisle, "It was a great relief to see eight figures coming back over the ridge".[71] Norman Fraser and Queenie Becker had reached the hut in a blinding snowstorm on the Saturday evening. They had sufficient food, there were dry blankets in the hut and, despite being unable to light the gas stove, they had survived quite comfortably for two days. Later that afternoon the frozen body of Mrs Kahn was discovered, lying in a curled-up position 400 yards from her husband. Allan and his fellow rescuers carried the three frozen bodies back to the farm where they were loaded into a waiting ambulance. It was a sobering lesson for all concerned. Mr JG du Toit, who had thrown open his farm to the rescue parties and earned their thanks, told the press that his farm would be closed to mountaineers in future. The UCT Mountain and Ski Club committee decided to abandon the hut extension scheme and to issue sheets of instructions and advice to leaders of future parties. Norman Fraser, while physically unharmed by his dreadful ordeal, was emotionally scarred for life.

One of the most spectacular and scenic locations, anywhere in the world, is the Cederberg, located 150 miles due north of Cape Town (Figure 1.6). It was the end of the academic year, December 1945, and Allan Cormack had just completed his Master of Science degree in physics. A party of six was assembled – Allan, Michael de Lisle, his sister Cecilia, Chrystal Komlosy, Walter Powrie and Elizabeth (or 'Ez') Coaton – and they spent ten idyllic days, from 13 to 22 December, hiking and climbing in an area rich in unusual geological formations, natural beauty and historical interest.[72]

The group set out on their hike from Citrusdal and, with a 75-mile slog ahead of them, decided to hire two donkeys – named Anthony and Cleopatra – to carry their food and camping equipment. However, it turned out that the animals were more trouble than they were worth.[73] As illustrated in the cartoon by Ez (Figure 2.8), the donkeys were extremely reluctant porters and Anthony was renamed Caligula in keeping with his character. Michael's encouragement was in Italian but the word *Sforza* used by Ez should actually have been *Forza*. The former is a well-known family from Milan while the latter means 'Move it!'[74] Despite the need to deal with two cantankerous pack animals, the six UCT friends had a memorable holiday, the details recalled with considerable pleasure sixty years later by Michael, Cecilia and Walter.

Highlights included climbing up through the Wolfberg Cracks, with huge red buttresses guarding the entrance, the peculiar rock formations sculpted by Mother Nature, and the impressive Wolfberg Arch, thirty feet high and an irresistible climbing challenge to the team (Figure 2.9). Also on their itinerary was an exploration of the Bushman Caves near Krom Rivier, which contained an enigmatic painting of an elephant hunt, estimated to be over a thousand years old, and the Raadsaal, a natural cavern formed by huge sandstone boulders, where the National Party was reputed to have gathered for meetings in the 1920s. While the Bushmen and elephants have disappeared from the area, the stunning Bushman painting endures (Figure 2.9).

In 1925 the UCT medical students had initiated the Hospital Rag – the word 'rag' apparently coming from the acronym for 'remember and give' – to raise funds for worthy charitable causes, particularly towards the support of those less fortunate citizens who eked out an existence on the Cape Flats. By the end of the war years the Students' Representative Council (SRC) had taken over the running of Rag and its main event, the procession of floats through the city of Cape Town and its primary thoroughfare, Adderley Street.[75] Although Allan was no longer a UCT student, it was not unusual for the junior staff to join in the festivities. He and his fellow mountaineers

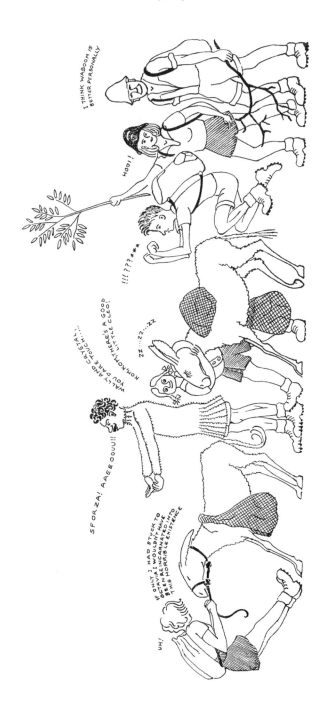

Figure 2.8 Cecilia de Lisle, Michael de Lisle, Ez Coaton, Allan Cormack, Chrystal Komlosy and Walter Powrie struggle with the donkeys, Anthony and Cleopatra, December 1945.

Figure 2.9   Wolfberg Arch (left) and a Bushman painting of an elephant hunt (right), 1945.

collaborated on the construction of a float with the theme 'Come landlord, fill the flowing bowl'.

After a week of hammering and painting, their float was taking shape, with a huge hessian-covered beer mug as the main feature. Unfortunately, someone had misappropriated their truck and the substitute, a slightly larger vehicle, could not quite accommodate the tavern which was designed to fit over the cab (Figure 2.10). Late on the Friday evening they stripped patches of Virginia creeper from the mathematics building to decorate the tavern and then early on Saturday morning, 23 March 1946, the floats assembled below the Mount Nelson Hotel on Orange Street. A party of twelve climbed onto the float including Allan, Michael de Lisle and his sister Cecilia, Emily Jarvis, Ez Coaton, Maurice and Hilary Norton, Jan Smit, and John Wanklyn, who "borrowed" a radar aerial from the physics department for the handle of the great beer mug. Smit produced 5 gallons of Vaaljapie – cheap white wine from Malmesbury – which guaranteed a non-stop party for the mountaineers.[76] The Rag celebrations became an important part of UCT's outreach to the community. By 1948 the annual amount raised was approaching £10,000, with a host of private welfare organisations as the primary beneficiaries.[77]

On the strength of his master's degree, Allan was awarded a two-year fellowship for overseas study. With the support of Professor James, he was accepted as a research student in the Cavendish Laboratory at Cambridge University, to begin his studies at the start of the academic term in October 1947. For the first half of that year he continued as a junior lecturer in physics at UCT.

Figure 2.10   The mountaineers and their float with the theme 'Come landlord, fill the flowing bowl' participate in the Rag procession, March 1946.

One of the highlights for South Africa was the Royal Tour of the country by the King and Queen, George VI and Elizabeth, accompanied by their daughters, the Princesses Elizabeth and Margaret Rose. On 22 April 1947 Allan, dressed in his academic robes and accompanied by his older sister Amy, was among the crowd of more than 500 who attended a special graduation ceremony in the Jamieson Hall. The Chancellor of the University of Cape Town and then Prime Minister of the country, Jan Christian Smuts, conferred the Honorary Degree of Doctor of Laws upon her Majesty the Queen.[78] After her speech, one of the few occasions during the Royal Tour when the Queen delivered a public address, the Chancellor concluded the ceremony with the solemn words:

> By virtue of the authority committed to me, I now dissolve this congregation of the University of Cape Town. God save the King.[79]

One of Allan's postgraduate colleagues in the physics department in 1946–47 was Aaron Klug. Klug was born in Lithuania in 1926 and grew up in the province of Natal, matriculating from Durban Boys' High School in 1942. At just sixteen, this precocious talent entered medical school at the University of Witwatersrand in Johannesburg but soon grew bored with medicine and switched to science, as his real interest lay in physics.

He had been drawn to Cape Town by the reputation of RW James, the world-renowned expert in X-ray crystallography, and by an advertisement for demonstrators, with a stipend of £20 per month. Having completed the coursework component of his MSc degree in 1946, Klug embarked on a thesis, the X-ray structure of a small organic molecule, that he published in *Nature* the following year.[80] The staff tea room was a lively place and it was there that Allan invited Klug to come climbing with him one weekend in the winter of 1947. Here, in Aaron Klug's own words, is a description of that day's climb, an experience he later recalled as "one final ascent":

> I used to be a kind of Sunday climber, if you could put it in that way, doing mild climbs on Kloof Nek, B climbs, occasionally C climbs, and Allan was an established climber. I didn't know that he had a route named after him, but he said he'd take me up something really interesting. I had done some climbing with ropes, but only a bit, just for short stretches, so he took me up Africa Face, which was a D or an E climb. About half way up, he was leading, above me, just the two of us. Suddenly I slipped off the ledge and swung out into the empty space and swung back again, hurting my arm. However, Allan held me, belayed the rope, pulled me up to safety and I was able to climb the rest of the way. I often think: What would have happened if he hadn't been able to hold me?[81]

Aaron Klug, who later became Director of the Molecular Biology Laboratory at Cambridge University, would go on to receive the Nobel Prize for Chemistry in 1982. The award was in recognition of his work on the structure of nucleic acid-protein complexes, realised in particular in viruses and also in chromatin, as revealed by electron microscopy.[82] The image of two future Nobel Laureates, hanging precariously by a hemp rope on the front face of Table Mountain, a few hundred feet above certain death, stretches the bounds of credulity.

Chapter 3

Physics and Friends at Cambridge

My interests had turned to nuclear physics, so of course I applied to Cambridge with visions of the Maxwell-Thomson-Rutherford tradition dancing in my head.

Allan Cormack[1]

It was almost eight centuries ago that a group of Oxford dons took exception to the hanging of two fellow scholars who had murdered a prostitute. They absconded to East Anglia, to the refuge of the Fens, where they founded a new seat of learning. The first Chancellor of Cambridge University was elected in 1225 and shortly thereafter, in 1233, he was recognised by the Pope, Gregory IX. Such recognition was not accorded Oxford University until 1254. "As so often", it has been said, "Cambridge started after Oxford and finished first".[2]

Medieval Cambridge was a commercial hub, with warehouses, loading docks and bars scattered throughout the town. There were numerous brothels, five of them reputedly squeezed between the Magdalene Bridge and the Church of the Holy Sepulchre, just three blocks up the road. The new university was certainly not greeted with any enthusiasm by the locals, and an uneasy standoff between town and gown ensued. That relationship survives to this day. Cambridge is no longer merely a cluster of ancient colleges, endowed by generous benefactors, with picturesque gothic buildings designed by architectural luminaries such as Christopher Wren. The city is now home to hundreds of hi-tech firms such as Microsoft and can rightfully claim to be the science capital of Europe.

Despite the juxtaposition of the old and the new, it is the characters who over the centuries have provided Cambridge University with its abiding charisma. The most famous of Cambridge's mathematicians was Isaac Newton, who developed both differential and integral calculus – later published as the *Principia* – and took up the Lucasian Chair of Mathematics in 1669 when he was just 27.[3] It is reported that the incumbent professor resigned his chair in favour of Newton since his pupil knew more than he did! There was the great classical scholar Richard Bentley, the master of Cambridge's largest and wealthiest college, Trinity, who in the early 1700s heaped scorn on the other fellows and dons, calling them "asses, fools and dogs". The Viennese philosopher Ludwig Wittgenstein, at a meeting of the Moral Sciences Club held at King's College in 1946, had an altercation with the visiting lecturer Karl Popper and threatened him with a poker! It reputedly required the intervention of Bertrand Russell, the mathematician and philosopher, to separate the protagonists.[4]

In the autumn of 1909 three young Cambridge undergraduates formed a life-long bond that had a significant influence, both directly and indirectly, on the life and development of Allan MacLeod Cormack. Reginald W James, son of a London umbrella maker, and W Lawrence Bragg, son of the physicist William H Bragg, studied physics, while James M Wordie, son of a Scottish carting contractor, majored in geology.[5] They all graduated in the summer of 1912. Bragg went on to join his father in Leeds to work in X-ray spectrometry, and just three years later they shared the Nobel Prize for their analysis of crystal structures. James and Wordie enlisted with Ernest Shackleton's ill-fated odyssey to the Antarctic, having to abandon the *Endurance* which was crushed by ice in November 1915, although their contributions – James as physicist and Wordie as geologist – led to the lives of the entire party of 28 being saved.[6]

By 1947, when Allan Cormack was contemplating postgraduate studies, Lawrence Bragg was Cavendish Professor of Physics at Cambridge, Jock Wordie was lecturing in geography at Cambridge and was a Fellow of St John's College, while Reginald 'Jimmy' James was, as indicated earlier, Professor of Physics at the University of Cape Town. Not only had James served as Allan's mentor, he was also an Old Johnian. It was therefore only natural that Allan was accepted by St John's College, where Wordie was appointed his tutor, and at the Cavendish Laboratory where, alas, Bragg was not to be his supervisor.

St John's College is situated at the north end of a single street that tactfully changes its name – from King's Parade to Trinity Street to St John's Street – so as not to offend the larger colleges. The college buildings straddle the River Cam (Figure 3.1) and are connected by the Bridge of Sighs. It was recently described by its master as "Not a fashionable college, but a great one".[7] The founder of St John's in 1511 was Lady Margaret Beaufort, the mother of Henry VII, whose motto was *Souvent me souvient.* Roughly translated this means 'Remember me often' and is captured in her floral emblem – incorporating marguerites and forget-me-nots – which frames the roseate brick gateway to the college.

Figure 3.1   St John's College chapel tower seen from the Backs across the River Cam.

For a young man who had just departed the far tip of Africa (Figure 3.2), St John's and Cambridge were a revelation: there were so many people, fellows, undergraduates and postgraduates, deeply interested in subjects that also interested him. Seldom did Allan lack the opportunity for lively discussions with people well versed in their fields. Occasionally this led to unintended consequences. Because South Africa's most eminent geologist Alex du Toit was a supporter of Alfred Wegener, Allan had grown up believing in Wegener's theory of continental drift.[8] It made sense to him because he knew that the botanical family of *proteacea*, found in abundance around Cape Town, occurred in only one other place on earth, in Western

Australia. When Allan mentioned the story of plate tectonics to the young geophysicists of St John's, they looked at him askance and fell silent. Some time later he discovered why. Their tutor was Harold Jeffreys. In addition to being one of the most distinguished dons at St John's, he was also violently opposed to the theory of continental drift. Because of his eminence, he was almost solely responsible for the rejection of Wegener's ideas until the 1960s.[9]

Figure 3.2   Amelia, Allan and Amy Cormack at Cape Town harbour just before Allan boarded the *Durban Castle* for his passage to Cambridge, September 1947.

Already engaged with their PhD studies in nuclear physics at the Cavendish were four other 'colonials' (as those from the dominions were called), each a few years older than Allan, who had all arrived in Cambridge a year earlier than him in the autumn of 1946. These included: another South African, Godfrey Stafford; two Canadians, Charlie Barnes and George Lindsey; and an Australian, Joan Freeman. During World War II Stafford had served in the South African Navy working on ship-borne radar, Barnes had participated in the Canadian contribution to the atomic energy project, first in Montreal and later at Chalk River, Lindsey had worked on radar and operational research for the Canadian Army, while Freeman contributed to the Radiophysics Laboratory near Sydney, developing radar for the defence of Australia.

In the years preceding the outbreak of World War II, the Cavendish Laboratory at Cambridge University was probably the most famous scientific institution in the world. But why was it so famous and what had inspired Allan and his fellow colonials to pursue their PhD research there? The surviving colonials – George Lindsey, Charlie Barnes and Godfrey Stafford – have provided some insight.[10]

The first Cavendish Professor of Experimental Physics was James Clerk Maxwell, who set up the laboratory in 1874. An outstanding mathematician, Maxwell produced the unified theory of electromagnetism and the kinetic theory of gases. He was followed by John William Strutt (Lord Rayleigh), who pursued research on sound, light, and electricity, supervised the expansion of the laboratory, and won the Nobel Prize for physics in 1904 for his work on gas densities and argon (Table 3.1). He was subsequently appointed Chancellor of Cambridge University (1908–1919).

The third Cavendish Professor was Joseph John Thomson, who held the post for thirty-five years (1884–1918). He was best known for his discovery of the electron, and won the Nobel Prize for physics in 1906 for his research on the conduction of electricity through gases. A measure of the quality and significance of the work at the Cavendish Laboratory preceding the First World War is that seven of Thomson's research assistants would eventually go on to receive Nobel Prizes (Table 3.1). After stepping down as the Cavendish Professor, Thomson served over twenty years as the master of Trinity College, Cambridge.

Ernest Rutherford was JJ Thomson's first research student, arriving at Cambridge from New Zealand in 1895. The most celebrated of his discoveries was made during nine years spent at McGill University, in Montreal, where he pioneered research on radioactivity and identified alpha- beta- and gamma-rays, followed by twelve years at Manchester where, accompanied by Geiger and Marsden, he employed the scattering of alpha particles to reveal the structure of atoms (Table 3.1).

Charles TR Wilson also joined the Cavendish in 1895 and studied the effects of electricity in moist and dry air. He invented the cloud chamber, enabling the paths of charged particles to be photographed by the tracks they made in supersaturated air (Table 3.1). Wilson was subsequently appointed to a professorship at Cambridge (1925–1934) where he continued to examine the effects of physics on meteorology.

During the First World War, Rutherford worked on acoustic methods for detection of submarines, returned to Cambridge in 1919 and succeeded his mentor Thomson to become the fourth Cavendish Professor. Soon after his return, Rutherford undertook a series of experiments using alpha particles emanating from a radioactive source to transmute stable nuclei by detaching a proton. He was aided by James Chadwick, who had studied under him at Manchester before travelling to Germany to work with Geiger, and had been interned there for the duration of the war. Chadwick conducted experiments which led to his discovery of the neutron (Table 3.1). Then, to extend his experiments beyond what could be done by bombarding nuclei with alpha particles, Chadwick recognised that it would be necessary to acquire a device able to accelerate charged particles to velocities well beyond those of the alpha particles emitted by radium. Chadwick would later play an important role in World War II in the cooperation between Britain and the United States on the development of the atom bomb.

After service in the Royal Field Artillery during the First World War, John Cockcroft studied electrical engineering at Manchester, graduated in mathematics from Cambridge, and then joined Rutherford's research team at the Cavendish in 1924. After the remarkable successes in achieving the transmutation of nuclei of light elements by bombarding them with the alpha particles from radium, Cockcroft shared Chadwick's desire for a device that could accelerate charged particles to a higher velocity. He joined forces with Ernest Walton, a doctoral student from Ireland, and in 1932 they designed a high-tension accelerator with a potential difference of over half a million volts. This proved sufficient to cause the disintegration of lithium-seven nuclei into two alpha particles. Rather than just driving a proton out of the target nucleus, the nucleus was disintegrated (or fissioned) into two equal parts. Cockcroft and Walton were the first scientists to split the atom (Table 3.1), thereby providing the necessary experimental evidence to confirm Einstein's famous equation, $E = mc^2$ (where E is energy, m is mass and c is the speed of light).

Patrick Blackett served as a naval officer in the First World War and then studied physics at Cambridge under Rutherford. In 1924 he extended Wilson's apparatus and obtained cloud chamber photographs of the transmutation of nuclei bombarded by alpha particles. With Giuseppe Occhialini, Blackett pioneered coincidence counting, which avoided the wastage of photographing unwanted events when searching for rarely occurring events such as the passage of cosmic rays (Table 3.1). He not only confirmed the existence of the positron, but also made important contributions to our knowledge of the earth's magnetic field.

W Lawrence Bragg was born in Australia, trained in mathematics at Adelaide and Cambridge, and then turned to physics in 1910. As indicated earlier, after graduating from Cambridge in 1912, Lawrence joined his father William Bragg who was professor of physics at Leeds University. Their experiments led to the development of X-ray crystallography for which they shared the Nobel Prize in physics just three years later (Table 3.1). In 1919 Lawrence Bragg moved to Manchester and in 1938 he succeeded Baron Rutherford of Nelson, who had died the previous year, to become the fifth Cavendish Professor.

The outstanding successes of the Cavendish Laboratory up until 1939 were characterized by their fundamental contributions to atomic physics. Many of the scientists had begun their training in mathematics rather than physics. While the experiments often required specialised equipment, it was possible to design this and have it produced in the lab's workshop without great expense. The technique for building such apparatus was often referred to as "sealing wax, string and half a gram of radium".[11]

For over sixty years the Cavendish Laboratory and its scientists had made extraordinary contributions to experimental physics, particularly in unravelling the secrets of the atomic nucleus. The discovery of fundamental subatomic particles such as the electron and neutron overlapped the brilliant theoretical expositions that emanated from across the English Channel. These included electromagnetic radiation and special relativity by the Dutch physicist Hendrik Lorentz, quantum theory by Max Planck in Berlin, and the remarkable and elegant theory of general relativity by Albert Einstein, who conducted his research in Switzerland and Germany. However, the spell of this 'golden age' of physics was irrevocably shattered by the onset of war in 1939.

During World War II the majority of the research scientists left Cambridge for projects related to the war effort. Many of the physicists became involved in the development of radar, a very important element for the defence of Great Britain, and later for the counter-offensive which brought victory. Then came the enormous Manhattan project to employ nuclear fission to create weapons with unprecedented explosive power. Chadwick played a key role in arranging the British contribution, most of which was undertaken in newly constructed laboratories in the United States.

When peace returned in 1945, Sir Lawrence Bragg faced a very different scientific world than had existed when he inherited the Cavendish in 1938.

Table 3.1   Nobel Laureates associated with the Cavendish Laboratory in the first half of the 20th Century (http://nobelprize.org).

Name	Field	Year	Accomplishment
John Strutt	Physics	1904	"For his investigations of the densities of the most important gases and for his discovery of argon in connection with these studies"
JJ Thomson	Physics	1906	"In recognition of the great merits of his theoretical and experimental investigations on the conduction of electricity by gases"
Ernest Rutherford	Chemistry	1908	"For his investigations into the disintegration of the elements, and the chemistry of radioactive substances"
Lawrence Bragg	Physics	1915	"For his services in the analysis of crystal structure by means of X-rays"
Charles Barkla	Physics	1917	"For his discovery of the characteristic Röntgen radiation of the elements"
Francis Aston	Chemistry	1922	"For his discovery, by means of his mass spectrograph, of isotopes, in a large number of non-radioactive elements, and for his enunciation of the whole-number rule"
Charles Wilson	Physics	1927	"For his method of making the paths of electrically charged particles visible by condensation of vapour"
Arthur Compton	Physics	1927	"For his discovery of the effect named after him"
Owen Richardson	Physics	1928	"For his work on the thermionic phenomenon and especially for the discovery of the law named after him"
James Chadwick	Physics	1935	"For the discovery of the neutron"
George Thomson	Physics	1937	"For his experimental discovery of the diffraction of electrons by crystals"
Edward Appleton	Physics	1947	"For his investigations of the physics of the upper atmosphere especially for the discovery of the so-called Appleton layer"
Patrick Blackett	Physics	1948	"For his development of the Wilson cloud chamber method, and his discoveries therewith in the fields of nuclear physics and cosmic radiation"
John Cockcroft and Ernest Walton	Physics	1951	"For their pioneer work on the transmutation of atomic nuclei by artificially accelerated atomic particles"

The pursuit of nuclear physics would never be the same again. The largest laboratories would be run by governments, not universities. Rigid security requirements would be enforced for the protection of information related to the design of nuclear weapons. There would however be important practical applications, quite apart from weaponry, in fields such as power generation and medicine.

The pre-war nuclear research group which had made the Cavendish Laboratory world famous had dispersed. Gone were the days when crucial discoveries in nuclear physics could be made with comparatively simple apparatus. The wartime search for nuclear weapons had required the creation of enormous reactors, enrichment plants, and other major investments far beyond the scale of what could be provided by a university in 1945. For post-war Cavendish, the most promising advances in front-line nuclear research appeared to be dependent on the bombardment of nuclear targets by charged particles accelerated by machines to velocities beyond what could be produced by the emissions of naturally radioactive substances.

Allan Cormack and his colleagues from the dominions had come to Cambridge to study nuclear physics. They were all disappointed with what they found. There was no Rutherford and it would seem that two key ingredients were lacking: leadership of the nuclear physics group; and ready access to top-class theorists. As Allan later recalled:

> It was not that there was nothing interesting happening in physics. The crystallographers who Bragg nurtured were working on the large organic molecules and were soon to produce results which revolutionised biology. Ratcliffe's group (including Ryle) was doing excellent work at the birth of radioastronomy, and low temperature physics was flourishing in the Mond Laboratory. Why then did we not shift to another field if we found nuclear physics unsatisfying? Partly it was intellectual arrogance (at least on my part) and partly it was the system. I wanted to study the fundamental interactions which held the nucleus together and I knew about crystallography but dismissed it as being 'just classical physics'. Unlike the system in the United States where a graduate student has a couple of years of taking coursework during which time he or she can find out what they want to specialise in, the system at Cambridge was that there was no required coursework and one plunged into a thesis topic. This made it hard to change fields without losing valuable time.[12]

It was X-ray crystallography that enabled Max Perutz and John

Kendrew in 1945, with Bragg's support, to solve the structure of haemoglobin, a protein that provides the red pigment of blood whose major function is the transport of oxygen from the lungs to the tissues. It was in 1953 at the Laboratory for Molecular Biology, created as an offshoot of the Cavendish Laboratory, that Francis Crick and James Watson conducted their pioneering research, concluding that the structure of DNA – deoxyribonucleic acid – was a double helix, thus unlocking the secret of genetic inheritance. Impressively, since 1958 when Fred Sanger won a Nobel Prize in chemistry for his work on the structure of insulin, twelve more scientists affiliated with the Laboratory for Molecular Biology – including Perutz, Kendrew, Crick, Watson and two South Africans, Aaron Klug and Sydney Brenner – would go on to win a Nobel Prize.[13]

And so it was that Allan, having learned the techniques of X-ray crystallography under RW James in Cape Town, passed up the opportunity to conduct his research under the supervision of Lawrence Bragg at Cambridge. Instead, as we shall see later, he chose to focus on nuclear physics, in particular the beta decay of helium-six.

For a few years immediately following the end of World War II, the normal composition of the Cambridge student body was distorted by the arrival of a contingent returning from wartime service. They brought personal experiences beyond what a high school or undergraduate education could offer. In the case of those coming to pursue postgraduate research in the physical sciences, some brought relevant experience from their wartime service. Both undergraduates and postgraduates were a few years older than would become usual for the intake of new students after some years of peace. Another deviation from student life before the war was that although Cambridge had not suffered heavy damage during the war, it was several years before Britain returned to the ready supplies of the services and materials (including food) that were normal in peace time.

Contact with fellow students at Cambridge occurred mostly at the colleges, which provided housing and meals for all undergraduates. In the post-war years there was insufficient accommodation in the colleges for all the postgraduate students, many of whom resided in boarding houses, but ate in their college. A significant feature of life for a research student was that he or she was expected to spend most of the year in Cambridge, whereas the undergraduates were only there during three ten-week terms, or trimesters, occupying less than eight months of the year. As a consequence, the college dining halls were full for less than two thirds of the year, leaving only postgraduate students and college fellows for the balance of the year. And, of course, the college fellows had their own very separate 'high table', with a superior menu, which was a significant perquisite in post-war Britain when rations were the norm.[14]

A large proportion of the research students at the Cavendish Laboratory had gone there immediately following their graduation from Cambridge with a bachelor's degree, and with no other scientific experience. Some came from other British universities, but Cambridge had always been very hospitable to research students with postgraduate degrees from outside Britain, especially from countries of the British Commonwealth, for whom a number of scholarships were available. In Allan's case, he was supported by a scholarship from the University of Cape Town, based on his first class honours result in his physics MSc.

In the early post-war years, intercontinental air travel was in its infancy, and the overseas research students did not have the time – or indeed the financial resources – to return to their home countries on ocean liners. Consequently, most of their personal contacts and friendships were made with their research colleagues, especially those in their own subject, or from their own college. There was also a strong attraction to the other fellow 'colonials' from the Commonwealth countries.

Among Allan's special friends from his University of Cape Town days was Godfrey Stafford, who was in Gonville and Caius College, just a few minutes' walk from St John's. Allan would wander over to his rooms in the evening and say "Come on, Godfrey, let's go for a walk".[15] Godfrey would put down his pen and they would walk long distances along the banks of the River Cam, talking about a range of topics, including their common calling: physics. Stafford remembers Allan as a great raconteur, able to expound on many subjects at great length and with a depth of knowledge.

The four original colonials – Stafford, Barnes, Lindsey and Freeman – discovered, as latecomers to the overstretched Nuclear Physics Department in the Cavendish, that they had limited access to supervisors and that there were few textbooks available to deal with their problems. They were conscious of being at a significant disadvantage compared to the students who had trained at Cambridge and who had had the benefit of excellent teachers during their undergraduate years. George Lindsey had the brainwave of organising regular get-togethers to discuss aspects of nuclear physics and he invited some of the English students to join them. It was hoped that the local students would inject their own knowledge into the discussions.[16]

This first meeting, held in one of the laboratories at the Cavendish, was a big disappointment to the colonials. There were too many people and the traditional English reserve meant that the local students were reluctant to contribute, particularly if they knew little about a topic. When they did contribute they expounded at length on a topic with which they were familiar at a level that was quite uninformative to the colonials. Lindsey declared the meeting a dead loss and suggested that they should rather wait for the meetings to die out naturally and then "after a decent interval, we should quietly start up a group for us four nit-wits, to get down to the real question of difficulties and problems".[17]

It was Freeman who conjured up the name, the Nuclear Nit-Wits Club. They agreed their club should be a secret one and that weekly meetings would take place not at the Cavendish but in the rooms of either Lindsey or Barnes, who were both in Emmanuel College. The success of the Nit-Wits Club stemmed from the enthusiasm and commitment that each brought to the enterprise. Apparently no one else in the Cavendish ever discovered their secret society. One more member was allowed to join in the autumn of 1947: Allan Cormack, who had just arrived from South Africa and who, like Stafford, had studied under RW James in Cape Town. Labouring under the same educational disadvantage as the founder members, Allan was considered eligible for membership.

For the first few meetings of the club, Lindsey brought along a recently published paper that dealt with an interesting and relevant topic in nuclear physics. With his sharp and perceptive mind he was able to highlight the important problems with which the team grappled until true understanding emerged. They worked through the paper carefully, pooling their knowledge, following the thread of the argument and, when necessary, identifying topics that required further background reading. Each person participated, taking turns to produce problems or discussing specific questions which had arisen in their own research projects. Each contributed their own special talents: Stafford, with his insistence on self-discipline and order; Barnes, with his perseverance until the problem was unravelled; Lindsey, with his wit and drive; Freeman, with her enthusiasm and sense of fun; and Allan, whose care-free manner belied an original mind yielding insightful comments (Figure 3.3).

The Nuclear Nit-Wits Club provided a substitute for the supervision and tutoring that the members found lacking in the post-war Cavendish. But it was much more than an opportunity for self-education. It also created a camaraderie, a special bond between the five 'nit-wits' that was to endure for the rest of their lives.

John Wanklyn was an undergraduate friend of Allan Cormack's at the

Figure 3.3 The members of the Nuclear Nit-Wits Club, Cavendish Laboratory, 1948-49. Top: Allan Cormack, Godfrey Stafford and George Lindsey. Bottom: Charlie Barnes and Joan Freeman.

University of Cape Town, where he had majored in industrial chemistry. Having started his career with the newly-established Council for Scientific Industrial Research in Pretoria shortly after World War II, he was sent by his employers in the autumn of 1946 to Cambridge University to learn about metallurgy and, in particular, corrosion.[18] In the first three months of 1947 Britain suffered the harshest winter for decades - weeks of deep snow and unbroken frost. Yet the terrible winter was followed by one of the best summers on record, with days of unbroken sunshine. During the summer break Wanklyn and Godfrey Stafford had enjoyed a holiday in the Western Highlands of Scotland, exploring the countryside on bicycles.

Wanklyn had been at Emmanuel College for a year when Allan arrived at Cambridge in the autumn of 1947. He invited his old friend to join him for a vacation over the Christmas and New Year break. Allan, who had last visited the land of his forbears as a 14-year-old in 1938, was pleased to accept the invitation. Anticipating bad weather and short days, they abandoned the idea of bicycles in favour of walking, hitch-hiking, and using trains and buses.[19]

North of Glasgow, while walking on the banks of Loch Leven in the valley of Glencoe, they met the minister from the local Presbyterian Church. Allan, who had developed a special interest in the history of Scotland, was delighted when the conversation turned to the famous Appin murder, a story that features prominently in Robert Louis Stevenson's *Kidnapped*. In the mid-eighteenth century, the Highlands had been oppressed by the English and the Campbell clan, which colluded with the English, was despised by the rest of the Highlanders.

In 1752, on the roadside not far from the place where Allan and Wanklyn were chatting to the minister, Colin Campbell had been shot and killed. The English authorities were quick to respond to the murder of the 'Red Fox' and they arrested James Stewart, a known enemy of the Campbells. 'James of the Glen' was carted off to Inverary, a town which served as headquarters for the Campbells, where he was thrust before a court that consisted mostly of the local clan. To no-one's surprise he was found guilty and sentenced to death. The English dragged him back to Ballachulish (Figure 1.2), on the banks of Loch Leven, where he was hanged on a grassy knoll just outside the village.

As the minister narrated the story to them, his voice quivered with emotion. He described the episode as "pure political terrorism" as if it had happened the previous week instead of 200 years before. He finished off his story with a poem he had written on the subject. He delivered the first two verses in English but then, lowering his voice, said "You'll no' understand the last verse but you must hear it in the Gaelic". He proceeded to recite the final verse, the words emerging in a slow rolling rhythm reminiscent of the Skye Boat Song. The two young South Africans were suitably impressed.

After their exposure to the long Highland memory they continued further north, hitching a ride to Fort William. From there they set off on foot, following a narrow road up towards Glen Nevis which led to the youth hostel. Although it was just two days before Christmas they were astonished to find the hostel surrounded by military vehicles. A mountain rescue team from the Royal Air Force (RAF) had been called out from their base on the northeast coast of Scotland to search for a man who had last been spotted a few days earlier. With inadequate clothing and poorly equipped, he had apparently set off to climb Ben Nevis (Figure 1.2).

Allan's climbing experience in the mountains of the Western Cape enabled the two students to volunteer their services and they arranged to join the RAF team in a search party the next day. One major benefit of their offer was access to the team's generous rations: as much tea, sugar, butter and bread as they wanted. Wanklyn well remembered the next morning, crouching over a huge pan, frying kippers for almost 40 men.[20] They participated in two searches, the second on Christmas Day. After being transported up the valley in jeeps, the men were divided into small groups. Allan and Wanklyn were assigned to Dusty Miller, a tough wee Scotsman from Glasgow whose favourite off-duty recreation was climbing and hiking in the mountains.

And so, 25 December 1947 found Allan, John and Dusty huddled in a narrow snow-filled gully, just beneath the summit of Ben Nevis, where they ate their Christmas lunch (Figure 3.4). Included in their rations was a can of 'self heating' soup: a narrow tubular section down the centre of the can was filled with fuel and, when lit, heated up the soup. Thus fortified, they continued thesearch after lunch but sadly, and perhaps predictably, the missing man was never found. A day later the RAF abandoned the search and the two friends were soon on their way, heading west towards the coastal town of Mallaig (Figure 1.2).

The railway journey from Fort William passed through beautiful country, stopping at Glenfinnan where the two young tourists were again exposed to another chapter in Scottish history. They visited a monument, commemorating the landing of Bonnie Prince Charlie in 1745. The Prince had been met there by clan leaders who supported his failed attempt to reclaim the English throne. His statue gazed wistfully from atop a tall tower,

Figure 3.4   Dusty Miller and John Wanklyn on Ben Nevis, Christmas Day, 1947.

a solitary brooding figure against the backdrop of a stormy sky. To the two friends the statue appeared to be a dark symbol of the prince's unsuccessful campaign.

That evening in Mallaig they boarded a ferry for Stornaway in the Outer Hebrides, passing through the Kyle of Lochalsh, the principal route of entry to the Isle of Skye from the Scottish mainland. As their boat entered the waters of the Minch, John leaned casually across to Allan and asked "You have the money, don't you?" Allan, undaunted, replied: "No, I have only five shillings".[21] The Post Office, their only source of funds, would be closed for several days over the New Year holiday period and so they would have to keep their wits about them.

On arrival in Stornaway (Figure 1.2), they checked in at the largest hotel near the harbour, explained their financial predicament to the proprietor who was very accommodating, and then settled into their rooms. Allan told John he was going down to the pub to have a beer and a chat with the locals. Within ten minutes he was back at the room, exclaiming to John, "It's useless – they are all speaking Gaelic!"

The following day was Hogmanay – New Year's Eve – and the hotel proprietor and his family invited Allan and John to go out 'first footing' with them. In this traditional custom, immediately after midnight the Scots go out to visit their friends. A symbolic gift, said to bring luck to the recipients for the forthcoming year, is a lump of coal. Since the effect is enhanced if the visiting party includes among them a dark-haired stranger, Allan was naturally recruited for this purpose and he was loaded up with coal.

The Hebrides are a stronghold of the conservative branch of the Presbyterian Church known as the 'Wee Frees', but on Hogmanay they abandoned their sense of probity. At each house the party visited the same ritual was followed: an exchange of greetings and then a toast was drunk. As the night wore on, the celebrations became more intense. Their last call was a small house into which a crowd of more than 30 revellers was squeezed. To the accompaniment of a man marching up and down the stairs while playing the bagpipes, Allan and John were introduced to the formal toast: in pairs, people raised their glasses before them and, after chanting a succession of Gaelic toasts, they drank their glasses to the bottom!

Their homeward ferry set sail well into the evening of 1 January 1948. Allan Cormack and John Wanklyn sat up on the deck in pitch darkness as the ship made its way back down the Minch which, on this occasion, fulfilled its reputation for rough seas. Understandably, the effects of the rolling of the deck were exacerbated by their celebrations earlier in the day! Within a few days they were back down in Glasgow, from where they endured an uncomfortable train journey – many Scots were travelling back south after the holiday break – back to Cambridge. As Wanklyn later recalled, "So ended a wonderful holiday remembered, 51 years later, as a series of tantalising blanks and vivid scenes".[22]

* * * * * *

Warren Abner Seavey was the Bussey Professor of Law at Harvard University and revered by his students: his lasting memorial encompassed the achievements of incessantly inquiring minds.[23] In the summer of 1948 his daughter Barbara Jean earned an MA in theoretical physics through Radcliffe, the women's college affiliated with Harvard. Without her parents' knowledge she applied to spend an academic year at Girton, one of the two women's colleges that were part of Cambridge University (the other was Newnham). Having been accepted for Part III of the Mathematics Tripos, Barbara obtained her parents' permission and then set sail from Boston for Southampton in early September.

On arrival in Cambridge, she checked into the Hermitage, a boarding house close to the downtown area, which was her base until Girton opened on 1 October.[24] Once Barbara had opened her account at the Midland Bank, been issued with a ration book and registered with the police, all formalities had been completed and she was ready to get to grips with the English and their customs. She was somewhat taken aback that tea, instead of coffee, was served for breakfast, passed up kippers in favour of porridge, and discovered that if she wanted toast, she had to hold it over the gas fireplace herself. It took her some time to understand the English money, particularly recognising the coins, appreciating their value, and getting used to the pre-decimal system (12 pennies to the shilling and 20 shillings to the pound). A favourite trick was to give the shopkeeper a coin she knew would cover the price of the item and let him calculate the change. Fairly soon, however, her purse was filled with small coins and she was obliged to consult her guide book to add up their value.

Girton College, established in 1869 as the first residential college for women, occupies spacious grounds about two-and-a-half miles northwest of the centre of Cambridge, the distance apparently serving as a deterrent during the Victorian era to visits from male students. A three-quarter hour walk to the university could be shortened to 15 minutes of leisurely riding with a bicycle. Barbara purchased herself a second-hand bike along with the necessary accoutrements: a basket, front and rear lamps, and a spare tube.

It did not take Barbara long to settle in. The dons invited the foreign students to have coffee after dinner and then two of the English girls gave a party for the group of Americans, of whom there were six: two from Bryn Mawr, one each from Cornell and Mount Holyoke, and two from Radcliffe: Barbara and, much to her surprise, since neither knew of the other's plans, Natalie Day.[25] It wasn't long before the four friends from Radcliffe and Bryn Mawr had made plans to cycle the 18 miles to the cathedral at Ely. Also known as Cromwell's Castle, it was established by William the Conqueror in the 11th century and is a masterpiece of medieval engineering. The American women were awed by this remarkable structure, which was unlike anything they had previously encountered in their own youthful country. The journey to 'the ship of the Fens' was a pleasant ride across the flat countryside with none of the friends suffering any after effects.

During the Michaelmas Term of 1948, Barbara had to select the courses for the Mathematical Tripos Part III, a series of advanced courses taken by the most talented undergraduates, culminating in three days of gruelling examinations in late spring. The word 'tripos' is derived from a 3-legged stool on which the examinees were traditionally required to sit while being grilled.[26] In course selection she was guided by her mathematics advisor Mary Cartwright and her physics advisor Bertha Jeffreys. Although both women had doctorates, Barbara referred to them as Miss Cartwright and Mrs Jeffreys, respectively.[27]

Mary Cartwright was a distinguished mathematician in her own right, making significant contributions to the theory of dynamical systems – how one state of a system evolves into another – and for which contribution she was elected a Fellow of the Royal Society in 1947.[28] Barbara Seavey knew that 'Miss' Cartwright was a scholar who had done much to raise the prestige of women in the academic world. When Dr Cartwright was lecturing at Princeton in the spring of 1949 and visited Harvard in Cambridge, Massachusetts, she arranged for her parents to meet with her renowned supervisor. Appointed Mistress of Girton later in 1949, Dr Cartwright continued her research work despite the administrative responsibilities and became the first woman to serve on the Council of the Royal Society. In 1969 she was honoured by Queen Elizabeth, becoming Dame Mary Cartwright.

The courses Barbara chose included 'Quantum Mechanics' taught by Paul Dirac, 'Theory of Numbers' by Robert Rankin, 'Statistical Thermodynamics' by Fred Hoyle, 'Electrodynamics' by Dr Powell and 'Probability' by Harold Jeffreys. The last-mentioned was Plumian Professor of Astronomy who had a poor reputation as a lecturer – his teaching style was said by one visiting student from America to be 'nonexistent'[29] – and he was married to Barbara's physics advisor, Dr Bertha Jeffreys. The list of lecturers was a veritable *Who's Who* of world-renowned authorities: Dirac, Nobel Laureate in 1933 for physics, which he shared with Erwin Schrödinger for "the discovery of new productive forms of atomic theory"; Rankin, an expert on prime numbers and devotee of the genius Ramanujan; and Hoyle, noted cosmologist whose theory of continuous creation of the universe was, for a time, a serious competitor to the 'Big Bang' theory.

Except for slight changes in notation and pronunciation, Barbara soon discovered that the lectures were very similar to those at Harvard. She judged the English students in physics and mathematics to be quite a way ahead of her and put that down to the earlier specialisation in high school. Barbara found out that a "practical" was what the English called a lab period. For her course in numerical analysis, she was surprised to discover that they used hand-powered machines for computing as opposed to the electrical ones she had used previously at Harvard.[30] The mechanical machines had no automatic multiplication and so she had to accomplish this by successive additions that were rather slow, not to mention frustrating, especially as she had been used to a more high-powered machine. Here at least the Americans were ahead!

One of Barbara's firm pals, taking many of the same courses as her, was a Welsh girl called Menna Jones. Between lectures they would spend their time doing the rounds of all the different places that served morning coffee.[31] One of their special haunts, facing King's College, was the Copper Kettle with its beautifully polished copperware in the shop window. Another favourite overlooked the market square with its stalls selling a wide variety of produce, from fruit and meat to antiques and books. Menna, who had been an undergraduate at Cambridge, took time to introduce Barbara to the winding passageways and hidden gems of the town. Such special knowledge of Cambridge would soon pay dividends when Allan entered her life.

Another good friend was Louise Moore with whom Barbara shared a wild but enjoyable trip to Oxford, where Louise's aunt lived.[32] With her sense of adventure, Barbara thought it would be a good opportunity to visit 'the other university' and, as the weather was pleasant, they set out by bicycle, allowing two days for the journey to Oxford. Their spirits were rather shattered when, about five miles out, they came across a sign informing them that it was 80 miles to Oxford! Undaunted, and in glorious sunshine, they pushed ahead. Despite Louise being slowed down by a heavy knapsack, the friends made good progress and at dusk they pulled over to spend the night at a charming country inn named the White Hart. Not far from Bedford, the friends were quite taken with the embellishments on either side of the fireplace in the dining room: a pair of doors taken from the cell where John Bunyan was imprisoned in 1675 when he wrote *The Pilgrim's Progress*. This added some local colour to a slap-up dinner – their first lamb chop since arriving in England – after which they soaked themselves in a hot bath and crawled into bed, exhausted.

The following morning they set out in a slight drizzle without being too downhearted. However, the drizzle soon turned into a soaking downpour, the wind whipped up, and not long thereafter they were pushing their bicycles. When Barbara and Louise encountered their first hill, they huddled together, discussed their options, and quickly decided to catch a train. They found the nearest railway station about 2 miles away and, after a 45-minute wait for the next train, they dried out. By this time, of course, the weather had cleared up but, because they had lost so much time, they continued on by rail. They climbed off the train just before Oxford and cycled the last ten miles, arriving at their final destination by a more direct route. Compared to the flat topography surrounding Cambridge, they were impressed with the rolling countryside around Oxford. After a few days with Miss Muriedas, an eccentric but conservative Spanish spinster, the friends made their way back to Cambridge by train.

From the time of the establishment of the Colleges of Girton (1869) and Newnham (1871), the women were permitted, under sufferance, to attend Cambridge University lectures and to sit the Tripos examinations but they were not admitted to degrees.[33] This discriminatory policy was changed in December 1947 and on Thursday 21 October 1948 the first woman's degree at Cambridge was conferred in the Senate House on Queen Elizabeth. There was great excitement in the two women's colleges, which she visited after the celebratory lunch at Christ's College.

In the dining hall of Girton, she had tea with a select gathering (scholarship students, postgraduates, alumnae, and overseas students) of which Barbara Seavey was a member, and was presented with a bouquet of flowers. Thereafter the Queen was taken on a brief tour of the college and passed along the corridor on which Barbara's room was located.[34] As she later reported to her parents, "You can now tell your friends that your daughter has spoken to the Queen of England!" On noticing the bowl of chrysanthemums, the Queen had asked whose room it was and Barbara, somewhat overawed by the occasion, nevertheless had the presence of mind to curtsy and answer, "It's mine, Your Majesty". Barbara was impressed that the Queen was able to carry off the performance in a light and natural manner. It wasn't long, however, before the red carpet had been rolled away, exposing the bare floor, and the women of Girton returned to their normal routines.

The Queen's first grandchild, Prince Charles, was born to her daughter Princess Elizabeth and Prince Phillip just a few weeks later, on 14 November 1948. The British people now had a new heir to the throne. In celebration of the Prince, Girton College served up a marvellous dinner of roast duck and toasted him with cider.

It was about this time in the Michaelmas calendar that the graduate and research students of St John's College arranged a party for those of Girton College. In a letter to her parents, Barbara Seavey made first mention of her new friend:

It was a most successful party and particularly so for me as I met a fairly pleasant character by the name of Allan Cormack.[35]

As it turned out, Dirac's course on quantum mechanics also played an important part in their story, for it was during these lectures that the 24-year-old South African from St John's College first became aware of the young American woman from Girton College.[36] Of course it is possible that Allan had something to do with arranging the get-together of the two colleges! After the party they were to spend quite a bit of time in each other's company.

With the approaching Christmas vacation (the last lectures for the term were held on Saturday 4 December), Barbara noticed a job advertisement on the college bulletin board: half a day of housework per week in return for room and board. This suited her well for she had no plans for the Christmas break and Girton students were obliged to vacate their rooms for the holiday period.[37] Besides, there was now a new person in her life and he too was not planning to be away over the break. In addition, she had not been putting enough time into her studies, with all the gadding about, and her conscience was beginning to bother her. So, on 14 December she moved to the home of Mrs Catherine Cobb at 38 Barton Road, Cambridge. Her responsibilities included bed-making and cleaning the upstairs bedrooms, some ironing and helping with the dishes after dinner. Professor Seavey and his wife were no doubt relieved to hear that a charwoman came in every day and took care of the messy work: cleaning the stoves, bathrooms and toilets! The Cobb family – husband and wife and four children under the age of ten – were pleasant people and Barbara was made to feel at home, particularly on Christmas Day.

With Allan nearby, life seemed a little less dull. One of the highlights was going to the pictures, where they watched *Lost Horizon*, a pre-war film directed by Frank Capra that told the story of a group of civilians whose plane crashed in the Himalayas and who were rescued by mysterious people from the valley of Shangri-la.[38] Then on Christmas Eve they attended a carol service at King's College Chapel. Apparently it was a "must do" outing but unfortunately everyone else in Cambridge seemed to have had the same idea. Despite the size of the chapel – with its vaulted ceilings, stone carvings and spectacular stained glass windows – it was jam-packed and the two students were so far back they could barely hear the service, although they did appreciate the choir's beautiful rendition of Hark the Herald Angels Sing. Between Christmas and New Year Allan also showed Barbara around the Cavendish Laboratory, which was one of the principal places she had wanted to visit. She described her experience to her parents:

> Allan is doing research in physics and, in particular, work on
> the Neutrino – a mysterious particle that is rather embarrassing
> to a physicist because they aren't sure it is a particle since the
> only known property it has is spin.[39]

On New Year's Eve 1948, Barbara Seavey and Allan Cormack had a "rather gay but somewhat strenuous" evening.[40] They joined a party of eight for dinner and thereafter moved on to the Dorothy Café, one of the few places in Cambridge that offered dancing. Since term was out, there were very few students around and the café was patronised by townspeople. At midnight, the band started playing *Auld Lang Syne* and everyone came thronging onto the dance floor, grasped hands and formed a large circle. They all joined in and sang:

> Should auld acquaintance be forgot
> and never brought to mind?
> Should auld acquaintance be forgot
> and days of auld lang syne?

For Allan and Barbara this was not so much a celebration of 'times gone by' (a translation of the ancient Scottish 'auld lang syne') but rather a portent of things to come. They danced the New Year in and, despite ruining a pair of nylon stockings and getting to bed late, Barbara was up and about the next morning, taking care of her duties in the Cobb household.

In neither Cape Town nor Boston had the postgraduates endured temperatures as cold as those of Cambridgein the winter of 1948–49. At her request, Barbara's parents had sent her a lovely warm bathrobe which was a lifesaver in the cold upstairs room at 38 Barton Road. By sitting as close as she possibly could to her little electric stove she was able to stay comfortably warm. However, her problems arose when it was time to go to sleep:

> I've never experienced anything as cold as getting into one of
> these beds. This cold damp atmosphere is most penetrating
> and gets into the bed when you make it in the morning and
> there it is waiting for you when you crawl in at night. I can
> now appreciate – still theoretically I'm afraid – the benefits of
> warming pans.[41]

By the second week of January 1949 Barbara was back at Girton, pleased to have more time for herself, and settling into the second ten-week term of lectures (Figure 3.5). She briefly considered applying for a Fulbright Scholarship, established by Senator J William Fulbright from Arkansas in 1946 as a much-needed vehicle for promoting "mutual understanding between the people of the United States and the people of other countries of the world".[42] This would have enabled her to study for a PhD, but she was deterred by the three-year time commitment. Besides, she now had another focus in her life.

Before the end of January Allan and Barbara were eagerly anticipating the May Balls/indexCormack, Barbara Jean (nee Seavey)!May Balls (which at Cambridge are held in early June!) and Barbara, recognising the tradition associated with these social events, asked her parents to send over the blue satin evening dress that she'd worn to a friend's wedding the previous summer in Boston.[43] In the meantime, the postgraduate students at Girton returned the favour and entertained their male colleagues from St John's. Delighted with the rapid improvement in their dancing skills, the young couple also continued to enjoy the pictures. One memorable film, costing just 30 pence each, was *It Happened One Night* starring Clark Gable and Claudette Colbert. This 1934 production, directed by the legendary Frank Capra, about a spoiled heiress running away from her family, prompted Barbara to exclaim, "I wish Hollywood could produce something as good as that now!"[44]

By late March their romance was in full bloom (Figure 3.6). Allan and Barbara took a beautiful afternoon off and walked along the River Cam to Grantchester, a few miles south of the Cambridge city centre.[45] The route was a paved public footpath running through the Grantchester meadow, a rural idyll considered to be amongst the best scenery within walking distance of Cambridge. It was a favourite stamping ground for the poet Rupert Brooke who in 1912 wrote:

> But Grantchester! Ah, Grantchester!
> There's peace and holy quiet there,
> Great clouds along pacific skies,
> And men and women with straight eyes,
> Lithe children lovelier than a dream,
> A bosky wood, a slumbrous stream,
> And little kindly winds that creep,
> Round twilight corners, half asleep.[46]

Figure 3.5   Barbara Seavey, an American student in Cambridge, 1949.

One of the special benefits for those studying at Cambridge was the opportunity to attend public lectures given by renowned authorities. On Tuesday morning 8 March 1949 Allan and Barbara attended a lecture by Wolfgang Pauli, one of the giants in quantum mechanics and fresh from the success of his Nobel Prize in December 1945 for his formulation of the Exclusion Principle.[47] His comments that day seemed incomprehensible to almost everyone with the possible exception of the academic staff. Among the exciting new advances he discussed was the 'renormalisation' technique in quantum electrodynamics – where individual particles can be considered from a distant viewpoint – and the pioneer identified by Pauli was a young professor at Harvard called Julian Schwinger. Barbara had been taught by Schwinger and so she was thrilled to realise that one of her mentors had been recognised as having great potential. In fact less than 20 years later, in 1965, for his "fundamental work in quantum electrodynamics, with deep-ploughing consequences for the physics of elementary particles", Schwinger would share the Nobel Prize with Sin-Itiro Tomonaga and Richard Feynman.

Another talented though eccentric speaker was Professor Dirac whose lecturing technique was to read verbatim from the galleys of his latest textbook, of which all the students had a copy.[48] On being chided about this method of delivery, he is reputed to have replied, "Why try to improve on what is perfect?"[49] Joan Freeman recalled this story about one of Dirac's public lectures at the Cavendish Laboratory:

> It was a formal talk in that there was a Chairman who first introduced the speaker. Dirac then proceeded to fill the blackboard with mathematics, addressing his personal unseen world in his characteristic fashion. When the lecture was over and Dirac had sat down, the Chairman expressed his thanks to the speaker.
>
> "Professor Dirac", he then asked, "would you be prepared to answer questions?"
>
> Dirac nodded and the Chairman looked expectantly at the audience. A young man near the back stood up.
>
> "Professor Dirac", he said, "I do not understand your fourth equation down on the left-hand side of the blackboard".
>
> Dirac sat impassively, staring into space as if he had not heard. A long silence ensued. The Chairman was becoming embarrassed. Finally he intervened.
>
> "Professor Dirac" he said diffidently, "are you prepared to answer that question?"
>
> "Oh", responded Dirac, suddenly coming to life and looking surprised, "was that a question? I thought it was just a statement of fact".[50]

A popular sporting tradition among Cambridge undergraduates, and taken most seriously, was a series of rowing races known as the 'bumps'.[51] The boats started out on the Cam at equal intervals and the idea was for each boat to try to catch and then bump the boat ahead. If the competitors made a bump that put them up a notch for the following race and the process continued until a series of races determined the boat that came out 'head of the river'. Barbara and Allan discovered the dangers of being a spectator as there was a group that kept pace with the boats along the riverbank. This was most easily accomplished on a bicycle, which resulted in a mad crowd of cyclists tearing along the banks, shouting encouragement for their particular college crew, and paying little heed to where they were headed! This frivolity culminated when the leading team had prevailed. Before World War II the boat that came out 'head of the river' would be burnt but that practice was abandoned in the post-war era and an old, defunct boat was substituted.

Figure 3.6   Allan Cormack, punting on the River Cam, 1949.

The other great rowing event in the spring calendar was the Oxford-Cambridge boat race, a 4.5 mile course on the River Thames from Putney to Mortlake in London. First contested in 1829, the race gave rise to each University's official sporting colours – dark blue for Oxford and light blue for Cambridge – and is now one of the most popular sporting events in Great Britain, with an estimated global television audience of 400 million. In March 1949, however, there was a far more modest audience with many supporters, including Allan and Barbara, listening to the radio.[52] It was a very close race that year: the boats were never more than a length and a half apart with Oxford leading all the way until the final few minutes of the 20-minute race. The commentator was shouting at the top of his voice and the listeners had great difficulty discerning which boat was in the lead as they approached the finishing line. In the end it was a memorable victory for the Cambridge team who edged ahead by just a quarter of a boat length.

For their spring break in the middle of April, Allan and Barbara planned a two-week trip to Paris.[53] Joining them were Frank Mercer, a friend of Allan's, plus Barbara's two American chums, Natalie Day and Louise Moore (Figure 3.7). After taking the train via London's Victoria Station to Ramsgate, they fortified themselves with sea-sickness pills before boarding the ferry. It was just as well. The ship was no sooner out of the harbour than a storm struck. A large wave came crashing onto the deck, soaking them all through to their underwear.

Then to add to their misery, an announcement over the loudspeaker informed them that the waves had damaged the ship and they were obliged to return to port. One consolation was that while most of their fellow passengers had been sick, the Cambridge students had escaped this condition. They had a choice of remaining on the ship for the night or travelling by train to Dover and catching the night ferry. The latter option appealed to them and so at midnight they set sail again, switching to the train at Calais around four in the morning, and arrived in Paris just before noon, exhausted. It was a trip they all vowed never to repeat.

Paris, they soon discovered, was well worth what they had to endure in getting there. Anticipating the superb French cuisine, their first excursion was to treat themselves to a steak dinner, the food being quite inexpensive by their standards. Aside from the traditional tourist attractions such as the Eiffel Tower, Versailles and the Cathedral at Chartres (the spectacular architecture dating from the same period as Ely Cathedral), they also went to see the opera *La Damnation de Faust* by Hector Berlioz. A full day was also spent wandering in the Louvre, where they all agreed the best exhibit, by far, was the sculpture 'Winged Victory'. Discovered on the island of Samothrace in the Aegean Sea, the marble statue of Nike – recognised as one of the finest artistic achievements of the Hellenistic Age (2nd century BC) – appeared to hover, wings outstretched, as it prepared to land.

Paris was crowded with tourists, and the Cambridge students found they were able to get by with sign language. Louise and Barbara were on their way to change money at an American Express bureau when they were approached by two Frenchmen who offered Louise a seemingly attractive exchange rate. The young Americans followed the two enterprising Frenchmen to a small café to arrange the exchange. However, after further discussion and haggling, the final offer was not that favourable and so Louise decided not to do business with them. A little later, at American Express, the official informed her that one of her $20 bills was counterfeit. He asked Louise where she had acquired the note and she claimed, at first, that the money had come from the United States. Then the penny dropped. In the small café she had let a $20 bill out of her hand for a second, long enough for one of the 'currency traders' to switch a counterfeit note. Louise had escaped unharmed, except for the loss of $20, and her fellow travellers decided the amusing story was almost worth the price! For Barbara and Allan, it was a timely lesson in the hazards of international travel.

Figure 3.7   Louise Moore, Natalie Day, Barbara Seavey and Allan Cormack returning from France, April 1949.

The final ten weeks of the academic year at Cambridge – the Easter Term – ran from late April to mid-June in 1949. It was a hectic period for Barbara, preparing for the Tripos examinations, and culminated in a successful May Ball when she and Allan enjoyed the company of their friends (Figure 3.8). Allan had suggested a cycling expedition around Normandy in France to his friend George Lindsey. Since Lindsey was from Canada he was able to converse in French and could therefore negotiate all transactions. The plan was to meet at Victoria Station in London. However when George arrived, he met not only Allan but also Barbara. Lindsey deduced that his role was to be not only that of a translator, but also that of a chaperone, a duty for which he had no prior experience![54]

Figure 3.8   May Ball 1949. Back row: Allan Cormack and Peter Sturrock; Middle row: Frank Mercer and Ralph Bick; Front row: Elizabeth Poyser, Barbara Seavey, Dzunia Bick.

The three young physicists took the boat from Southampton, landing at Le Havre in Upper Normandy, and mounted their bicycles as soon as the formality of customs was over.[55] Their first night found them in a charming village called Honfleur, which was the seaport from which Samuel de Champlain first departed in 1603 for his explorations of Canada. One of Lindsay's tasks, when seeking accommodation in French farmhouses, was to explain that they wanted one double bedroom for Allan and George, and one single bedroom for Barbara, rather than their best room, which could easily have accommodated all three! A room with running water cost just a dollar a night, while a mouth-watering meal, including wine and coffee, came to between one and two dollars. At their first dinner, not realising it was a three-course meal, they over-indulged on the *Hors d'Oeuvres*, thinking it was the main course. When the main course was served, they had to pass on it and move straight on to the dessert. Thereafter they learned the value of gastronomical patience!

Their summer vacation was a leisurely two weeks spent cycling from Le

Havre to St Malo, a total distance of about 200 miles. On their journey they passed through the towns of Caen, Bayeux, St Lô, Granville and Mont St Michel. Each morning they were awakened by the noise of reconstruction – many of the villages had been badly damaged during the war when General Eisenhower's forces landed in Normandy on D-Day, 6 June 1944. Just five years later the French people with whom they associated were particularly friendly towards the South African, American and Canadian, remembering the role played by the Allied troops in their liberation. When the group returned to England they caught a ferry from St Malo.

As the ship made its way north past the Channel Islands of Jersey and Guernsey, Barbara and Allan were no doubt contemplating the future and what it held for them. For Barbara, her parents had insisted that she return to the USA. On 21 July 1949 she set sail on the Queen Mary. For Allan, the immediate challenge was to complete the research towards his doctorate in nuclear physics.

* * * * * *

To penetrate the nucleus, Cockcroft and Walton had built a voltage multiplier circuit in which they devised an intricate stack of capacitors linked together by a series of rectifying diodes that operated as switches. By opening and closing the switches in the appropriate sequence, they were able to convert a transformer potential of 200 kilovolts into a final potential of 800 kilovolts. In this way they accelerated protons down a 2.5 metre long vacuum tube and succeeded in releasing the two alpha particles from the lithium nucleus.

After Rutherford's death in 1937, the Cavendish contracted Philips of Eindhoven in the Netherlands to manufacture two linear accelerators based on the Cockcroft-Walton design (Figure 3.9). When Joan Freeman arrived at the Cavendish in October 1946, she was given a guided tour of the Laboratory by Dr Sam Devons, one of the senior research staff. She later recalled her first impressions of the accelerator hall:

> The most vivid memory I have is my first sight of the two high-voltage machines. They stood in awesome immensity in a huge hall, which must have been about fifty feet high and perhaps even more in length. They looked like imaginative ultra-modern sculptures. Each had two pairs of slender brown insulating columns, carrying many circular fins and large, flattened, regularly spaced doughnuts of silvery-grey metal. The doughnuts

were linked by a zig-zag lattice which rose to a huge bun-shaped metal terminal, somewhere up near the roof, surmounting and dominating the whole structure. The overall appearance of the design, in its elegant proportions, was beautiful – the more so in fact, for being entirely functional – and it made a dramatic and unforgettable impact on me.[56]

Figure 3.9   The Cavendish HT1 Cockcroft-Walton linear accelerator, 1949.

The smaller of the two accelerators, known as HT1 (the HT standing for 'high tension'), had been bought in 1937 and had seen important wartime service. It was capable of reaching voltages slightly in excess of a million volts. The larger, more recently acquired accelerator was known as HT2 and was capable of reaching voltage levels between 1.5 and 2 million. Each machine stood on top of its own control room, located at opposite ends of the HT hall.

Dr Bill Burcham was in charge of the HT1 generator and under his supervision was Joan Freeman.[57] Dr Devons was the leader of the group that worked on HT2, and he looked after the Canadian, George Lindsey. Godfrey Stafford and Charlie Barnes were engaged in two different parts of the same experiment with Professor Otto Frisch as their supervisor. Although theoretically capable of reaching 2 million volts, HT2 was somewhat unreliable and suffered frequent breakdowns. When Allan Cormack arrived at the Cavendish in 1947 he was assigned to a third group that was dependent on a cyclotron to accelerate its bombarding particles. The cyclotron had an interesting genesis.

Prior to the success achieved by Cockcroft and Walton, a young physicist at the University of California in Berkeley conceived a radical idea. Ernest O Lawrence recognised that while the linear accelerator proved useful for heavy ions such as mercury, lighter projectiles such as alpha particles required a vacuum tube many metres in length. This voltage multiplier system was limited to about 2 million volts because of electrostatic breakdown. Lawrence judged this design to be impractical. Instead, he imagined a design in which a magnetic field was used to bend the particles into a circular path so that they might pass repeatedly through the same electrode.[58]

In concept the cyclotron was an ingeniously simple device. It consisted of a flat round vacuum can, into which were inserted, back-to-back, two semicircular electrodes shaped like the letter 'D'. The device was then inserted between the poles of a powerful magnet and protons were introduced into the centre of the can. As the voltage switched back and forth, the protons began to move in a spiral, gradually moving outwards from the centre at greater and greater speed. By the time they reached the edge of the can and were hurled through a window, the protons had acquired a huge amount of energy. The cyclotron thus operated on the principle of resonant acceleration, although the effects of relativity became problematic at higher energies, requiring synchronisation to keep the accelerating frequency in phase with the relativistic mass increase.

By August 1931 Lawrence's cyclotron, with a diameter of just 28 centimetres and an electrode potential of only 10,000 volts, was able to accelerate protons to an energy level of 1.1 million electron volts. Although Cockcroft and Walton had managed to beat Lawrence in the race to split the atom by just a few crucial months, his revolutionary design changed the course of modern physics and he was recognised with the award of the Nobel Prize in 1939.

James Chadwick, having discovered the neutron in 1932, the same year that Cockcroft and Walton conducted their auspicious experiment, was keen to build more powerful accelerators at the Cavendish. However, Rutherford was reluctant to ask the authorities for the necessary funding and refused

Chadwick's request. So, in 1935 Chadwick accepted the chair of physics at Liverpool University where he soon set about the construction of a cyclotron with a diameter of 94 centimetres. He sought the assistance of Cockroft for the design of the magnet for his cyclotron. Rutherford was furious with Chadwick for this perceived indiscretion and the two never spoke again, thus bringing to a sad end their long friendship.[59]

After the death of the parsimonious Rutherford, Cockroft took responsibility for the design of a cyclotron for the Cavendish Laboratory that was completed just prior to the outbreak of World War II (Figure 3.10). It was on this machine – later described by Cavendish Professor Brian Pippard as a 'modest cyclotron'[60] – that Allan conducted the first experiments of his PhD thesis, the measurement of the beta-spectrum of helium-six (He^6). However, the 'modest cyclotron' proved to be notoriously unreliable, and it seemed to Allan that the second half of his thesis, begun in early 1949, would take years to complete. Despite frustrations with the temperamental instrument, his sense of humour remained intact:

> The cyclotron energy had been raised so that my original supply of pure He^6 was now badly contaminated by other gaseous beta-emitting isotopes. The circumstances which caused this are amusing. Television was broadcast only in the evening, except for special events. Reception in Cambridge was never very good (the transmitter was 50 miles away) and sometimes it was very bad, the screen being covered with horizontal streaks. This was the case on the morning of an Oxford-Cambridge boat race when many students were gathered around the television sets. At least one viewer knew that the cyclotron oscillator radiated an harmonic which interfered with the TV signal. A peremptory call to "turn off the damned cyclotron" resulted in an immediate improvement in the picture and subsequent permanent increase in the frequency and energy of the cyclotron.[61]

Although John Cockroft had been appointed to the Jacksonian Chair in Experimental Physics in 1939, his wartime responsibilities intervened. When the war ended, Cockroft returned to the UK from Canada and was persuaded to take up the directorship of the newly-established Atomic Energy Research Establishment at Harwell, in Oxfordshire. His commitments at this new national facility precluded him from taking up the Chair, although he encouraged collaboration between his scientists and those at the Cavendish. Towards the end of 1947 Cambridge University offered the vacant Jacksonian Chair to Otto Frisch and he took up the appointment early the following year.

Figure 3.10   The Cavendish's 'modest cyclotron', used by Allan Cormack for his research on helium-six, 1949.

Frisch came to the Cavendish with a well-established track record. Born and educated in Vienna, he suffered the fate of many Jewish scientists during the Nazi era and was forced to flee Germany in 1933. Having worked with Niels Bohr in Copenhagen for almost five years, late in 1938 he visited his famous aunt, Lise Meitner, in Stockholm.[62] In Sweden they received the news that Otto Hahn and Fritz Strassman had bombarded a uranium nucleus with neutrons, with one of the byproducts being barium. Meitner and Frisch hypothesised that the uranium nucleus had been split into two more or less equal parts and that a large amount of energy – which they were able to calculate – had been released. They called the process 'nuclear fission' and Frisch, with characteristic humour, referred to the byproducts as 'fission chips'.[63] Hahn and Strassman were awarded the Nobel Prize for chemistry in 1944, while Meitner and her young nephew were considered by many unfortunate not to have received similar recognition for their elegant contribution to physics.

While the postgraduate students at the Cavendish respected Frisch for his intellectual status and his ability to design original scientific apparatus, it was clear to them that he lacked the necessary leadership qualities to regain the momentum of the nuclear physics group that had been lost with the interlude of the war. Charlie Barnes, who had been allocated Frisch as his supervisor, was struggling with his project.[64] After some time he went to see Frisch and explained the problem. Frisch listened for a minute or two and then suddenly asked "Who is your supervisor?" "Well, you are Sir", answered Barnes, and Frisch, with no hint of embarrassment, responded, "Oh, am I?"

Robin Cherry, a physics lecturer at the University of Cape Town, would follow Stafford and Cormack to the Cavendish a few years later, in September 1954, and he experienced the same problem as Barnes. Cherry was sent off by Frisch to conduct the necessary background research for his PhD when, after three months, he discovered that someone else had already reported on the work during the previous summer at a conference. When Cherry reported this to his supervisor, Frisch responded, "Oh, but I was at that Glasgow conference when the material was presented. I must have been asleep".[65]

Despite the extraordinary legacy of Thomson and Rutherford, the Cavendish Laboratory could not recapture its pre-war glory in nuclear physics without a dynamic leader. While Frisch was not Allan's main supervisor, it is clear that his lack of interest in each student's progress was a significant impediment to those students who were struggling with their projects. How differently things might have turned out if Allan had chosen to work on X-ray crystallography with Lawrence Bragg as his supervisor.

Stafford, Barnes, Lindsey, and Freeman completed successful research programmes which had depended on the high tension generators HT1 and HT2. All four obtained their Cambridge PhDs. After a temporary return to South Africa, Godfrey Stafford became the Director of the Rutherford Laboratory, Britain's national centre for high energy and elementary particle physics research, and later Master of St Cross College at Oxford. He is a Fellow of the Royal Society.

After teaching at the University of British Columbia for four years, Charlie Barnes accepted an appointment at the California Institute of Technology, where he served as a professor of physics until his retirement in 1992. During this time he and his postgraduate students collaborated frequently with William A Fowler on studies of the nuclear-astrophysical synthesis of the chemical elements within stars. For his pioneering studies on this subject Fowler was awarded a Nobel Prize for physics in 1983.

George Lindsey returned to Canada, where he became the Chief of the

Operational Research and Analysis Establishment in the Department of National Defence. He was subsequently made an Officer of the Order of Canada.

Joan Freeman remained in England, and enjoyed a long career at the Atomic Energy Research Establishment at Harwell. She shared the Rutherford Medal with Roger Blin-Stoyle for her experimental contributions to the theory of beta-decay. Her autobiography *A Passion for Physics: The Story of a Woman Physicist* includes several chapters describing life in the Cavendish in the post-war years.[66]

Years afterwards, at a meeting of the American Physical Society, Allan Cormack and a fellow physics professor from Tufts University, Brenton Stearns, bumped into Sam Devons. Stearns later recalled the incident:

> Allan seemed to appreciate the irony in many situations, but with humour where others might have been bitter. In particular, he once told me about the collapse of his doctoral research. He had been working on the details of a beta decay scheme that was crucial in understanding the coupling after parity violation had been discovered (memory says it was He^6). A man called Allen at the University of Illinois had just published a similar result. Allan had then said, "Then there was nothing to do but bugger off". This was made clear to him by his research director, or laboratory head, Sam Devons. I thought it rather strange, because every result should be independently checked. Then Allan said, with a hearty laugh, "It turned out the bugger got it wrong!"[67]

With the unreliable performance of the cyclotron, and little helpful supervision, Allan was unable to make sufficient progress with his research. By the autumn of 1949, with time running out, and his financial resources almost exhausted, he was in a difficult predicament. As George Lindsey recalled: "When Barbara Seavey returned to the States, Allan's normal happy demeanour descended into a depressed melancholy". His fellow Nuclear Nit-Wits put this down to the capricious behaviour of the Cavendish cyclotron.[68] Meanwhile, five thousand miles away in Cambridge, Massachusetts, Professor and Mrs Seavey found a similar misery in their returned daughter. Warren Seavey was a man of action, so he invited Allan to Boston for Christmas.

Allan wanted to marry Barbara Seavey but in those days one did not marry without a steady job. So he wrote to his old mentor, Professor Reginald James, on the off-chance that there might be a vacancy at the University of Cape Town. James replied immediately with a telegram offering Allan a lectureship. In fact, he almost did not get the message but, thanks to the efficiency of the porters at St John's College, the telegram was read to him over the 'phone in the office of the stationmaster in Liverpool Street Station in London where he was intercepted on his way to the vacation in the United States.[69]

Allan Cormack and Barbara Seavey were married in Boston on 6 January 1950 and later that same day they flew from Logan Airport to the United Kingdom. And so, in February 1950, Allan returned to Cape Town with a bride but no PhD, and, since there was no cyclotron in South Africa, pursued no further work on helium-six.[70]

Chapter 4

Return to the Fairest Cape

So I started to think about the problem *ab initio*. It did not
take me long to realise that the problem was a mathematical
one.

Allan Cormack[1]

Allan and Barbara spent the first few weeks of married life in the United
Kingdom. Having accepted the lecturing post at the University of Cape
Town (UCT), Allan bade St John's College, the Cavendish Laboratory and
Cambridge University farewell, and in early February 1950 he and Barbara
boarded a ship at Southampton. The two-week sea voyage to Cape Town,
during which Allan celebrated his 26th birthday, was an ideal opportunity
for them to relax and contemplate what lay ahead.

Allan was optimistic about his career in nuclear physics, about taking
up residence at his *alma mater* and rejoining his mentor, Professor RW
James. He was also excited at the prospect of introducing Barbara to his
family and friends in South Africa and sharing with her his love of the wild
mountains of the Western Cape. While he had been studying at Cambridge,
there had been a major shift in the political landscape of his home country:
in 1948 the Nationalist Party, under DF Malan, had come to power and
subsequently introduced apartheid, the policy by which the population was
segregated along racial lines. It was into this politically hostile climate that
Allan and Barbara would be settling: physics and family would become
something of a refuge.

The last leg of their sea voyage brought them down the west coast of
Africa. Then, without warning, the unmistakable silhouette of Table Moun-

tain appeared. Allan's spirits soared as he recalled the many times he had pitted his climbing skills against the obstacles: Sheerness, Union Route and Procrastination Crag. As their ship threaded its way between Blaauberg Strand on the mainland and Robben Island to enter Table Bay, he pointed out to Barbara the different routes up the front face of his beloved mountain. Her thoughts were elsewhere, however, contemplating what lay ahead for her on this strange continent.[2] She need not have worried: there was a friendly reception awaiting them as they disembarked at the docks. There to meet them on the quay were Allan's mother Amelia, his sister Amy and Mr William Foster, mayor of Cape Town and the father-in-law of his brother Bill (Figure 4.1).

Figure 4.1   Mr William Foster, Mrs Amelia Cormack, Barbara and Allan Cormack (left to right) at Cape Town docks, February 1950.

Allan assumed duty as a lecturer in physics on 10 March 1950, replacing Dr RW Guelke, a specialist in electro-acoustics, who had succeeded Brian Goodlet as professor and chairman of electrical engineering.[3] Allan's contract, signed by the chairman of the UCT Council, William Duncan Baxter, was a concise two-page document that succinctly laid out his conditions of employment, which included a three-year period of probation.[4] The starting salary for all lecturers was £550 per annum, increasing by £25 each year up to a maximum of £800, which was the entry-level salary for a senior lecturer. Allan's promotion to senior lecturer would be dependent upon the needs of the department, a distinguished record in research, and a long and outstanding service in teaching or administration. Professors, of whom there were relatively few (since UCT followed the British tradition of only one or sometimes two professors per department), enjoyed considerably better remuneration than their junior colleagues, with salaries that ranged between £1,000 and £1,300 per annum.

The principals of South Africa's universities were deeply concerned about the finances of their institutions, and in particular about the remuneration of their staff.[5] They issued a statement urging the government to appoint a commission to investigate the situation. In the statement they pointed out that in 1930 the annual state subsidy had been £55 per student. Over the following twenty years the number of students had trebled but the student-based subsidy had remained unchanged, despite increases in the cost of living. Dr EG Malherbe, principal of the University of Natal, revealed that the proportion of their annual operating budget covered by the state subsidy was less than 40 percent, but it was deemed impossible to raise the already high student fees. He pointed out that young BSc graduates from Natal received higher salaries than their former lecturers, while the emolument of a trawler captain was almost double that of a professor. The response from the Secretary for Education was that no increase would be possible as the total grant had been pegged for 1949–1951. However, the Holloway Commission on University Finances and Salaries was established and its subsequent report in mid-1953 led to some relief and an increase in salaries for lecturing staff.[6]

In the physics department, Professor James was glad to have Allan back, with or without the PhD. In fact James himself did not have a PhD and many of his generation did not set much stock by them.[7] Allan renewed his friendship with fellow lecturer John Juritz and settled in to a comfortable relationship with the two senior lecturers in the department, Drs Eric Grindley and Walter Schaffer. Also in the physics department was one of the Cormack family's friends, George Laing, who was from Mey in Caithness. George was a senior mechanical technician in the departmental workshop and would later play a vital role in some of Allan's research. Allan was asked to teach subjects which in larger institutions would have been reserved for lecturers more expert than him. As he commented to

Professor James at the beginning of his second year in April 1951:

> Not having the dynamics course has enabled me to do more
> with the second year MSc students and I might now be able to
> finish off with something about the quantum theory of fields – a
> subject about which I should like to know some of the details![8]

Shortly after his arrival at UCT, Allan was approached by Marinus van den Ende, the Wernher and Beit Professor of Bacteriology in the university's medical school. Professor van den Ende was a medical graduate of UCT who had earned his PhD at Cambridge University just as the hostilities of World War II broke out.[9] He immediately joined the Royal Army Medical Corps where he gained personal recognition for his contribution to the fight against typhus and the development of a vaccine. He returned to UCT in 1946 to take up the Chair and quickly set about the creation of a research programme with a focus on virology, in particular the study of bacteriophages and rickettsias. With his boundless energy, and his international publications on topics such as bovine lumpy-skin disease and the bacteriophages of *pseudomonas aeroginosa*,[10] van den Ende soon attracted the attention of the Council for Scientific and Industrial Research (CSIR). The CSIR provided the necessary funding for the establishment of the Virus Research Unit in 1950, with the indefatigable van den Ende as its inaugural director.[11]

Professor van den Ende asked Allan if he would like to try and put the infection of bacteria by bacteriophages on a quantitative basis. Allan thought this would be an enjoyable project and so he accepted the challenge.[12] Assisting with the research and providing many helpful discussions were Professor F Holliman, an organic chemist, and Dr A Polson. The latter was a biophysicist who had spent time in the Laboratory of Physical Biology at the National Institutes of Health in the USA, working on the diffusion constants of the bacteriophage called *Escherichia coli*.[13] Polson had returned to the Onderstepoort Veterinary Research Laboratories in Pretoria before van den Ende managed to recruit him to Cape Town to join the Virus Research Unit. Polson was recognised as a scientist of exceptional ability and had become internationally renowned for his work on particle size.[14]

When a liquid culture of bacteria is infected with bateriophage, the phage particles attach themselves to the cells, infecting them. There were three principal parameters to be considered here: the concentrations of viable cells, infected cells and the free phage particles, respectively. Based on the theory of Brownian motion, Allan set up three differential equations governing the rate of change of the concentrations. Linking these three equations were three constants: the multiplication of viable cells according to an exponential law; the probability of an uninfected bacterium being infected; and the probability of an infected bacterium being multiply infected. Values for these constants were calculated using experimental data provided by van den Ende, Holliman and Polson. Allan used standard numerical techniques to solve the differential equations and, on the basis of his theory and the data generated by the medical scientists at the Virus Research Unit, he wrote the first draft of a manuscript.

Believing that he needed an independent opinion, Allan sent a copy of the draft paper to Professor EE Briggs, in the School of Botany at Cambridge University. Allan said that he had studied at Cambridge and mentioned their brief acquaintance through a postgraduate student, Frank Mercer. He was looking for affirmation that he had produced a piece of scholarly work that might lead to publication, and commented to Briggs:

> As you will see, the treatment gets a bit mathematical – far too mathematical for Professor van den Ende who claims his mind goes blank as soon as he sees an integral sign. I am continuing the work but don't know whether I am wasting my time or not. As there is nobody in Cape Town who knows anything about the more mathematical side of biology, I am writing to you as a person who does, to get an opinion as to whether it is worth going on with the work or not.[15]

Briggs was encouraging although he did comment that Allan should consider gathering better quality data as this would definitely strengthen the impact of his theory. By mid-1952 Allan had completed two manuscripts, parts I and II, with the title 'A quantitative treatment of the infection of bacteria by bacteriophages'. The first of these papers was sent to Dr AFB Standfast of the Lister Institute of Preventive Medicine in England who was the editor of the *Journal of General Microbiology*.[16] Unfortunately the paper was rejected and so Allan was discouraged from submitting part II for consideration. It is unclear whether the reason for the paper being turned down was because the mathematical theory was beyond the grasp of bacteriologists but Allan was undoubtedly discouraged from pursuing this line of research.

Ten years later, when he was on the brink of throwing out the old manuscripts, Allan contacted Salvador Luria, Professor of Biology at the Massachusetts Institute of Technology.[17] Luria would later share the Nobel

Prize in Physiology or Medicine with Max Delbrück and Alfred Hershey for "their discoveries concerning the replication mechanism and the genetic structure of viruses".[18] While Luria believed the theory to be valid and the assumptions reasonable, he did not think Allan's calculations would find sufficient utility to justify publication except perhaps as part of a broader, experimental paper.[19] And so it was that Allan's first attempt to apply his mathematical skills to a medical problem never saw the light of day.

The year 1952 was an auspicious one for many white South Africans since it was the tercentenary of the landing at the Cape by the Dutch explorer Jan van Riebeeck in April 1652. It was also the Jubilee year of theSouth African Association for the Advancement of Science, since it had been founded in Cape Town in 1902. The Mother City was the fitting choice to host the celebration and the Council of UCT offered their facilities at Rondebosch for the business of the meeting.[20] The Association's congress met under the Presidency of Dr Basil Schonland from Monday 7th to Saturday 12th July. The generosity of the government of the Union of South Africa made it possible for the organizers to invite distinguished scientists from overseas.[21] For Allan and his colleagues in the physics department at UCT – RW James, John Juritz and Colin 'Boot' Butler – the congress provided an ideal opportunity to learn about the latest developments in their field from some of the world's leading scholars.[22]

From the United Kingdom came Sir Edward Salisbury, Director of Kew Gardens, who delivered a Popular Evening Lecture on the gardens, and Sir Lawrence Bragg, Director of the Cavendish Laboratory. Belgium sent Canon Georges Lemaître, Professor of Astronomy and Mathematics at the University of Louvain, who had formulated the theory of the expanding universe, while from France came the famed palaeontologist, Dr Henri Vallois, Director of the Musée de l'Homme in Paris. Holland sent Professor LEJ Brouwer, founder of the Intuitionist School of Mathematics, while Sir Kerr Grant, the distinguished physicist from Sydney, represented the Australian and New Zealand Association for the Advancement of Science. Professor James was the representative of the Royal Society of South Africa, an organisation whose mandate was the publication of original scientific work.

In his Presidential Address, Schonland reviewed the history of the Association, from its founding in 1902, when the Anglo-Boer War was drawing to a close, and he also considered the future of science in South Africa. He reminded the delegates of the objects of their Association as articulated by the founding President, Sir David Gill, His Majesty's Astronomer at the Cape:

To set forth the claims of science to the sympathy and support of every citizen, to give a stronger impulse and a more systematic direction to scientific enquiry, to promote the intercourse of societies and individuals, to obtain a more general attention to the objects of pure and applied science, and to remove any disadvantages which may impede its progress.[23]

Allan was particularly interested in the presentation by Lemaître, who spoke about the clusters of nebulae in the expanding universe, and whose ideas had been adopted and expanded by Sir Arthur Eddington. Another notable contribution by Lemaître in the field of cosmology had been his hypothesis of a 'primaeval ato' from which the universe was derived.[24] For James it was an opportunity to renew his friendship with Bragg, his old chief at Manchester (Figure 4.2), who provided a fascinating overview of the current research being conducted at the Cavendish Laboratory.

Figure 4.2   Professor RW James and Sir Lawrence Bragg at the University of Cape Town, July 1952.

There were in all some 140 researchers in the Laboratory, of whom 100 were PhD students while the remainder were research fellows and scholars. In addition, there were 45 technical staff members who assisted with the research and built the specialised experimental apparatus. The Laboratory was divided up into a number of sections, including nuclear physics, radio waves, low temperature physics, meteorology, electron microscopy, crys-

tallography and biophysics. Bragg, an Australian by birth, was a strong
advocate of the need to recruit talented postgraduate students from the
Commonwealth and other foreign countries, bringing fresh ideas and valuable contacts. A policy that he pursued with vigour was to give his research
units the independence to determine their own research agendas, believing,
"It is penny wise and pound foolish, however, to institute any system of
checks which occupies the scientist at his office desk and keeps him away
from his laboratory and study".[25]

Since research in nuclear physics required the use of very large and
costly apparatus, the Cavendish Laboratory was still limited by the two
high-tension sets, the cyclotron and the van de Graaff machine that Allan had encountered, and been frustrated by, in the late 1940s. Although
Bragg and his colleagues were planning to build a linear accelerator capable of producing particles with energies up to 300 million volts, they had in
the meantime developed their own molecular beam method for measuring
nuclear spins and moments, thus breaking the American monopoly on this
line of research. Pioneered by Isidor Rabi of Columbia University in the
1930s, molecular beam technology would later play a critical role in the
development of magnetic resonance imaging in medicine.[26]

Bragg also described some of the work in low temperature physics, in
particular the enigma of superconductivity, a curious phenomenon that
was to have an enormous practical impact four decades later. The electron microscopy group had developed a novel technique for using X-rays
to capture a pair of stereoscopic images with a spatial resolution of less
than one micron. He projected slides of the fly *Drosophila*, showing all its
internal organs in their correct relative positions in three dimensions in a
very striking way. Bragg's own special interest was the use of X-ray crystallography to understand the complex structure of solid state substances.
He highlighted the work of Max Perutz, who was concentrating on the con-
stituents of living matter, particularly haemoglobin, and described some of
the challenges they faced in analysing the data:

> Another branch of the work is concerned with methods of com-
> puting results. X-ray analysis has a character all its own. The
> experiments are really simple; no mathematics of a high order
> need be used in deducing the results. The difficulty lies in the
> very great complications of the analysis. It is rather like break-
> ing down an extremely hard cypher. Methods of doing heavy
> computations rapidly and without strain are an essential part
> of the investigation. We use the punched card system, an elec-

tronic digital computer, and optical methods of simulating the X-ray results.[27]

Despite Allan's determination to pursue a career in nuclear physics, the juxtaposition of the two reports on the use of X-rays – the three-dimensional images of *Drosophila* and the mathematical and computational analysis of protein structure – must have taken root deep in his subconscious scientific mind.

It had been more than five years since Allan had published his last peer-reviewed journal article, the results of his own master's thesis which had appeared in the *Transactions of the Geological Society of South Africa.* Although he had supervised a few MSc theses since his return to Cape Town from Cambridge, these were on instructive if not fundamental topics and did not encompass sufficient originality to warrant publication. Allan urged his students to go overseas to experience what he called the "real thing".[28] One of his students was Robin Cherry, who later became professor of physics and dean of the science faculty at UCT, who used a scintillation counter that Allan had built – one of the first in South Africa – to measure the angular correlations in cobalt 60.[29] Cherry held his supervisor and their department head in high regard:

> I was one of the small third-year physics class in 1952 which had the extraordinary good fortune to be lectured to by both James and Cormack. An unforgettable combination they were, too: James a lecturer of outstanding clarity, shy and kindly, already an academic of stature, respected and revered by us all; Cormack lively and enthusiastic, amusing and irreverent, friendly and helpful.[30]

Allan's drought of publications was broken in 1953 with a paper on momentum determination from cloud-chamber photographs that appeared in *The Review of Scientific Instruments.* In this brief technical note he demonstrated how it was possible to determine the radius of curvature of a charged particle in a non-uniform magnetic field; the only requirement was for the field to be symmetrical about the axis of the cloud-chamber, enabling the use of much greater magnetic fields.[31] During the following two years, in a sudden burst of productivity, he published a further five papers, all of them theoretical and all appearing in a prestigious American journal, *The Physical Review.* He explored two primary themes: the scattering of gamma rays by the nucleus and atomic electrons; and heat generation in the earth by neutrinos from the sun.

Allan considered three eponymous types of gamma ray scattering –
Thomson, Rayleigh and Compton – which referred to the different processes
by which gamma rays could interact with nuclei and atomic electrons. He
outlined separate methods for the study of high-energy gamma ray tran-
sitions, arguing that such an approach was necessary because of the short
lives of the states involved.[32] In his other line of research, Allan showed how
the carbon cycle in the sun caused a large flux of neutrinos at the earth.
He calculated the heat generated in the earth as a function of the neutrino
magnetic moment, showing that it was small compared to that resulting
from natural radioactivity that emanated from the earth's core. He further
demonstrated that practically all of the carbon cycle energy liberated in
the form of neutrinos was lost by the sun and that their energy distribution
when reaching the earth was the same as at the time of their creation in
the sun.[33]

Stanley Amdurer, an American ex-serviceman, came to UCT in the
early 1950s to begin postgraduate studies in geology, arriving from the
USA with his South African-born wife Frieda Neugebauer, a civil rights
activist.[34] In addition to the funding he received on the GI Bill, he also
managed to secure an appointment as a temporary lecturer in the geology
department. Since his research focus was in engineering geology, specifically
the study of clay, Amdurer needed an X-ray diffraction unit. The equipment was prohibitively expensive, however, and well beyond the reach of
an impecunious research student. Fortunately for the American, Allan be-
friended him. Allan, who had used similar equipment for his master's thesis
eight years before, built an X-ray diffraction unit in the basement of the
physics department, scrounging many of the parts and using his ingenuity to manufacture the rest.[35] For Stanley, this equipment was a lifesaver,
enabling him to begin the research that would lead to his doctorate.

The friendship between the young South African physicist and his geological counterpart from the USA grew and they soon turned their scientific talents to other projects. The nuclear industry was growing rapidly in
the early 1950s and uranium-bearing minerals were in demand. Allan and
Stanley decided to conduct some prospecting on wheels. They rented an
old fashioned limousine, a large Buick that had a glass partition between
the passenger section in the back and the driver. They mounted a dozen
Geiger tubes, each 3 foot long, on the roof and connected these to the
pen recorder that was propped up between the front passenger seat and
the dashboard. An array of 12 volt wet-cell batteries was installed on the
floor in the rear of the vehicle where Stanley was seated, his feet squeezed

between the batteries.

The aspirant prospectors had an idea that the most promising place to look for uranium was in Namaqualand, an area close to the border with South-West Africa (Figure 1.6). After an all-day journey from Cape Town, they arrived at their destination and set up camp. Allan was the driver while Stanley sat in the back with the responsibility for navigating and recording the headings of their traverses. They criss-crossed Namaqualand, hoping to pick up significant signals. As the Buick negotiated the rough roads (Figure 4.3), the batteries bounced around in unison, splattering acid over Stanley's shoes, socks and blue jeans. Their perseverance was rewarded, however, and by the third day they were recording signals that were off the scale. Excitedly they gathered samples over a large area of almost 100 square miles. Allan and Stanley convinced themselves they were rich! However, on their return to UCT, when Amdurer analysed the rock samples back in his laboratory, he discovered they had been recording the radioactive particles of thorium, a mineral with little value. There was no uranium. With Stanley's clothes in tatters and his ankles burned from the battery acid, the prospectors decided it was time to retire gracefully.

Another of their joint projects involved the testing of a seismic device that had been designed for shallow depth investigations by Ian Gough of the CSIR. At the time deep surveys were performed using explosives. With this new device they placed a flat steel plate on the ground and then struck the centre of the plate with a ten pound sledge hammer. A set of highly sensitive 'geophones' was used to record the seismic waves. When the device was first tested, during daylight hours one weekend, they picked up the noise of the distant traffic. Their next strategy was to work in the quiet of the night, setting up their apparatus on the Cape Flats, a sandy bush-covered area just east of town. Again Allan and Stanley recorded tremendous background noise, which puzzled them. After some investigation, they heard a low-pitched braying sound and discovered the source of the noise: a restless donkey, tethered to a tree. They spent the entire night gathering further data, although they established that their hammer blows were unable to generate very effective seismic waves. Clearly they were ahead of their time because today portable geophysical devices are in widespread use.

When Allan and Barbara first arrived in Cape Town they moved in temporarily with his mother and sister at 15 Brays Court in Station Road, Mowbray. It was not long, however, before they rented their own flat in Pinelands, at 30 Morningside in Forest Drive. Although their new home was a few miles further away from UCT, the Cormacks had by now ac-

Figure 4.3   Prospecting in Namaqualand with Stanley Amdurer, 1953.

quired their own motor car, a Morris Minor, which they drove to campus. Barbara had accepted a post as a junior lecturer in the mathematics department, teaching first and second year students.[36] Weekends were spent exploring the Cape Peninsula and enjoying the pristine beaches at Fish Hoek, Llandudno and Hout Bay.

While Barbara was not exactly enamoured with Allan's passion for climbing, she did join him for a hike up to the UCT hut at Waaihoek, and even managed to scale the summit herself (Figure 4.4). Walter Powrie, Allan's great climbing companion from their high school and undergraduate

days, had married Ez Coaton and the four of them drove up to the Cederberg for a weekend getaway in December 1951.[37] The two couples made
their base camp at Algeria on the Friday evening, setting out early the
following morning when it was still relatively cool. By the time they had
hiked up to Uitkyk Pass, the temperature had risen dramatically and was
approaching 40° Celsius. Barbara suffered heat stroke, which gave Allan,
Walter and Ez cause for concern. When the party came off the mountain
later in the afternoon, they packed up the car and returned to Cape Town
so that Barbara could be treated. Fortunately she soon recovered but the
experience dampened any enthusiasm she might have had for hiking and
climbing.

Figure 4.4   Barbara and Allan at the summit of Waaihoek Mountain, 1952.

Despite the demands of his job at UCT, Allan still found the time
to continue climbing. Among his regular companions were two of Cape
Town's leading climbers, Mike Mamacos and Colin 'Boot' Butler, the latter
a senior undergraduate whom Allan had first met in 1947 before he left
for Cambridge.[38] A favourite climb was the Zeebasberg, in particular the
Gingerbread Face on the Grey Buttress (Figure 4.5). Nearby was the UCT
hut at Waaihoek where, in the winter months of July and August when
sufficient snow had fallen, they turned their attention to skiing. None of

Figure 4.5   The Zeebasberg (left) with Mike Mamacos scaling the Gingerbread Face (right), 1952.

them was an accomplished skier but whatever they lacked in ability was more than compensated for by enthusiasm: after carefully waxing their pre-war skis, Allan, Mike and Boot would climb up above the hut and launch themselves downhill. Seldom were they able to complete the 300-yard run without taking a tumble in the virgin snow.

In March 1952 Allan and Barbara moved into one of the UCT staff houses, 5 Lovers' Walk, Rosebank, just beneath the main campus. Their monthly rental was £16.13.4 and occupation was given for a year at a time up to a maximum of three years.[39] A regular visitor to their new home was Boot, who was completing his degree in physics that year and also teaching part-time at Observatory Boys' High School.[40] He and Allan would stay up late, covering the full spectrum of science teaching, from high school to university, and chatting about their other common interest: cosmology. Barbara indulged them in their long after-dinner discussions, as they argued over the latest hypotheses of Hermann Bondi and Fred Hoyle and, of course, the current politics in South Africa.

In August of that year, Barbara gave birth to the Cormacks' first child, a daughter whom they named Margaret (Figure 1.2). Among the frequent visitors to the staff houses to see the new arrival was Allan's mother, his older sister Amy and her husband-to-be, Leonard Read. Len, who had served as a pilot in World War II and joined UCT as a young clerk from the Auditor-General's Office in 1946, was working in the university's administration department.[41] Amy had been employed in the Jagger library at UCT when a post became vacant in the administration office. The Deputy Registrar Mr PG McDonald had asked her if she would like to fill it and so "I went to work for Len and have worked for him ever since!"[42]

By the time Margaret was old enough to walk, Allan and Barbara would take her to Kirstenbosch Gardens on the slopes of Table Mountain, to the beaches at Camps Bay and further afield to Hermanus. Shortly before her third birthday, in July 1955, they packed up the family car and set off on their own 'Great Trek', a 1300-mile journey to the north, to the world-renowned Kruger National Park. The Cormacks spent their nights in thatched huts in the camps at Skukuza and Punda Maria and explored the park during the day. It was the first time Barbara had seen giraffes, hippos, elephants, lions and the other animals in their natural habitat. This exposure to the wilds of Africa left a lasting impression on the young American mother.[43]

With their three-year lease on the UCT staff house coming to an end, it was again time for Allan and Barbara to find another home. They decided to rent a house in Pinelands, at 4 Mead Way, so they could be closer to Allan's mother and also to Len and Amy, all of whom were now living in Pinelands. Christmas 1955 was a happy occasion for the Cormack clan, with Allan's older brother Bill, his wife Marjorie and their son Ian driving down from Johannesburg (Figure 4.6). Barbara and Amy were both pregnant: Jean Cormack was born on 5th January 1956 (Figure 1.2) and her cousin Amelia Read, known as 'Jo', arrived four months later. It was to be another 25 years before the Cormack family would be together again.

* * * * * *

In September 1950 the UCT community mourned the death of Field Marshal Jan Christiaan Smuts, who had been Chancellor of the university since 1937. He had also been installed as the Chancellor of Cambridge University when Allan was studying at the Cavendish Laboratory. Smuts had earned a double first in the law tripos at Cambridge in the early 1890s and was regarded by some in the academic community as a "king among students" and a person whose intellectual spirit and guiding influence had left an indelible mark on UCT.[44] The university chose Judge A van de Sandt Centlivres, Chief Justice of the Union of South Africa, to succeed Smuts as Chancellor. He was installed on 6 April 1951 in an induction ceremony attended by, among others, Dr DF Malan, Prime Minister of the Union

Figure 4.6   A Cormack family reunion, December 1955. Back, left to right: Allan; his
mother Amelia; his brother Bill and wife Marjorie; William Foster. Front, left to right:
Ian Cormack; Barbara and daughter Margaret; Amy and Len Read.

who was also the Chancellor of Stellenbosch University. In his speech, the
UCT alumnus Centlivres paid tribute to General Smuts and then stated
his own position and that of the University of Cape Town regarding their
relationship with the apartheid government:

> There can be no progress without this academic freedom. It
> is the lifeblood of a university. The fact that the universities
> are subsidised by the state does not give the state the right of
> interfering in the universities' affairs.[45]

By the early 1950s, about 6 percent of UCT's 5,000 students were black.
Although the university was proud of its tradition of accepting students
on the basis of academic merit alone, when Non-European students – as
they were then called – arrived on campus they discovered a two-tiered
system in place: academically they were fully integrated into the student
body but socially they were prevented from playing sport for the university
or living in a residence. The 'colour issue' divided both the Students'
Representative Council (SRC) and the wider student body itself. Mr Vic

Wessels, a member of the SRC and classified 'Coloured' by the apartheid laws, asked:

> The SRC is elected by all students and is responsible to all students. Can it refuse to defend the right of a Non-European to use the swimming pool?[46]

It would be another five years before the student body and its leadership was ready to embrace the principle of social equality. The actions of the government contributed to this change of heart. When two Africans, also referred to as 'Natives' at the time, were elected to the SRC in September 1953, parliament's House of Assembly debated the position of the 'open' universities, Wits and UCT.[47] A Nationalist Party backbencher said that educated Natives became agitators and the African population was not civilised enough to absorb a university education. Dr Thomas B Davie, a Stellenbosch graduate who had been dean of medicine at Liverpool before his appointment as Principal and Vice-Chancellor of UCT in 1948, made the university's position quite clear:

> Academic freedom consists of four fundamental freedoms. These are the freedom of the university to determine without outside pressure or force what to teach, whom to teach, how to teach, and who shall teach. It is only by fighting that liberty was ever attained or maintained.[48]

The government responded by setting up a three-man commission, under the chairmanship of Dr JE Holloway, whose mandate and terms of reference clearly emphasised the primacy of apartheid principles in university education.[49] The South African Police Force were the custodians of apartheid and did not hesitate to enter the campuses to enforce the country's laws. One of Allan Cormack's students was an Indian man from Pretoria who was arrested in his laboratory but, because some people from the previous regime were still in positions of influence, the student was back on campus a week later.[50]

Support for UCT came from various quarters. Clive Gray, then President of Student Government at the University of Chicago described the United States Supreme Court's landmark ruling that 'separate but equal' educational facilities provided for under-privileged racial groups were in fact disgracefully inferior;[51] James Pickett, President of the Scottish Union of Students, expressed solidarity, condemning the threat to introduce apartheid into all South African universities; and Mr B Ngwenya, Chairman of the Cape branch of the African National Congress, who was in-

vited to speak on campus, described the upcoming People's Congress and
encouraged UCT and the National Union of South African Students to
participate.[52] Professor RW James had been appointed acting Principal,
following the premature death of Dr Davie. Since James was a shy and
retiring man, Allan and his colleagues in the physics department wondered
how he would acquit himself in the political cauldron. They need not have
worried.[53] In a protest meeting at the City Hall, James attacked the government propaganda with his customary clarity and elegance:

> We believe academic segregation to be wrong in principle and
> wasteful and inefficient in application. To build and staff new
> colleges would be a costly matter and would seem inadvisable
> at a time when all our universities need greater financial sup-
> port for the scientific and industrial development of the country.
> But our main objection is on the grounds of principle and not
> expediency.[54]

Groote Schuur Hospital, the main teaching hospital of the UCT medical school, had been built just prior to the outbreak of World War II on
land that once belonged to Cecil John Rhodes, in the shadow of Table
Mountain (Figure 2.1). Among the many special services offered was radiology, the department being divided into two sections: radio-diagnosis and
radio-therapy.[55] As the volume of work increased, so did the complexity of
the apparatus and its consequent maintenance. The need for a physicist
was first recognised in the early 1940s but, given the financial stringencies
of the time, no appointment was made. As an interim solution, Professor RW James was appointed honorary consulting physicist and was later
followed by Dr R Guelke of the physics department. Shortly after the
hostilities ended, a full-time physicist was recruited from England: Mr D
Wakely, who had earned an honours degree in physics from the University
of Manchester.[56]

In the latter half of 1955 Wakely resigned, ostensibly because he was
"fed up with the way the bloody doctors are treating me", and took a job
in Canada.[57] The hospital was suddenly faced with a crisis because by then
South African law required a qualified nuclear physicist to supervise the
clinical administration of radioactive isotopes.[58] Without such supervision
the radio-isotope programme, and the treatment of patients, would have
come to a halt. With Wakely gone, Allan was the only person in Cape
Town who had been trained in nuclear physics and the handling of radio-
isotopes and so he was prevailed upon to help out. He agreed to spend a
day and a half per week – Fridays and Saturday mornings – for which he

was remunerated. It was a decision that would change his life forever.

Allan went to work at Groote Schuur Hospital on 2 January 1956, with a commitment to continue until the end of June the same year. Since there was as yet no department of nuclear medicine, he was placed in the radiology department, where one of his responsibilities was to look after the film-badge service. These badges, worn on their white coats by the X-ray technicians (or radiographers as they were also known), measured the amount of radiation accumulated during the week. Allan was required to read the individual badges and then to interpret the measurements as an accumulated dose for each technician. He had not been working very long at the hospital when he noted a curious practice. On a Friday afternoon, before going home, the radiographers would take their coats off and hang them next to the safe where all the radium was kept.[59] The badges would then accumulate a significant dose over the weekend. Since the total dose determined how many extra days of vacation each worker received, Allan soon stopped the practice, moving the coats when he came in to read the badges on Saturday mornings. Although he was very unpopular when he first introduced the policy, the radiographers and clinicians soon recognised and respected his talents as a caring and thoughtful colleague.

The head of the department was Dr James Muir Grieve, a Glasgow-educated radiotherapist, who introduced Allan to the treatment of malignant tumours with X-rays. Ideally, Grieve wanted to deliver a known radiation dose to the tumour while simultaneously delivering a small dose to the healthy, surrounding tissue. He explained to Allan the current practice for treatment planning in his own department, which closely followed the methods used in the United States, Great Britain and Sweden.[60] The planning was based on the use of isodose charts that were obtained for X-ray beams passing through homogeneous materials, even though Grieve acknowledged that the human body was highly inhomogeneous.

With his training in the physical sciences, Allan was surprised at the crudeness of the method. It occurred to him that in order to deliver an accurate dose to a specific point in the body, one had to know the absorption coefficient for X-rays at all points between the skin surface and the point of interest. He further realised that this information had to be obtained from observations made with equipment that was exterior to the body. Grieve suggested the necessity of taking measurements through a range of different angles. It was immediately evident to Allan that the problem was a mathematical one.[61]

Before he could spend any time trying to solve the problem, however,

Allan had a more pressing commitment: he and his family were preparing to go on sabbatical. For the Cormacks, the United States was clearly the place to go.

Chapter 5

A New Beginning in Boston

My wife, who had spent most her life in suburban Boston, had
come with me to the wilds of Africa without demur. It was only
fair to reciprocate by returning with her to what, for me, were
the wilds of America.

Allan Cormack[1]

Originally, when he first applied for furlough and study leave in July 1955,
Allan had proposed spending a year at Cornell University in Ithaca, upstate New York, participating in their nuclear research programme.[2] However, since Barbara's family lived near Boston and she had studied physics
at Radcliffe, the Cormacks decided that Harvard University would be a
better choice for the sabbatical year. Barbara knew several of the faculty
members in the physics department from coursework and advising in the
mid-1940s. She recommended that Allan write to Norman Ramsey, which
he did. Ramsey sent back a friendly reply, inviting Allan to work at the
Harvard Cyclotron Laboratory.

In 1956 the mighty dollar reigned supreme and so the Cormacks faced a
serious financial problem. Fortuitously, Allan managed to secure an overseas scholarship from the CSIR which, together with his UCT salary and
the termination of the lease on their house in Mead Way, Pinelands, meant
that the family was able to afford its first overseas sabbatical. In early July,
Allan and Barbara, together with their two young daughters – four-year-
old Margaret and six-month-old Jean – boarded the ship for Southampton,
from where they travelled by a second boat to Boston.

When the Cormacks arrived in the USA they were met by Barbara's parents, Warren and Stella Seavey, and her brother Bob and sister-in-law Dorothy at the Boston harbour. Professor Seavey had retired from Harvard in July the previous year, aged 75, having taught the law of torts and agency for 49 years.[3] He had had a long but incredibly itinerant career as an academic lawyer, with stints at Pei Yang University in China (1906–1911), and then at Oklahoma, Louisiana, Indiana, Nebraska and Pennsylvania, before finally returning in 1927 to his *alma mater*, Harvard Law School, where he remained until retirement.

Warren Seavey had had only a brief interaction with his son-in-law in late 1949, when Allan came to Boston to marry Barbara. However, the two academics immediately established a positive rapport, assisted no doubt by the young South African's insistence on drawing up a pre-nuptial agreement, giving Barbara control over her own property.[4] (In South Africa at that time, a wife's property belonged to her husband.) Although retired, Warren Seavey was still in demand as a visiting lecturer at universities where his former students now taught. The Cambridge apartment that he and Stella kept close to Harvard was therefore often empty, and so they offered it to their daughter and her family. The Cormacks were thus fortunate to have an ideal home base for their sabbatical year.

The apartment was within walking distance from the campus of Harvard University and the Cyclotron Laboratory, which was located between Oxford Street and Divinity Avenue (Figure 5.1). Allan's collaborators were Norman Ramsey, Richard 'Dick' Wilson and Joe Palmieri. Ramsey had been the director of the lab from 1948 to 1950 and, although his main research interest lay in molecular beams, he had a paternal interest in the cyclotron, having helped to design and build it. He was present at 2:03 am on 3 June 1949 when protons were first extracted from it.[5] Wilson was a young Englishman with a doctorate from Oxford who had just been recruited to Harvard from the Atomic Energy Research Establishment at Harwell in the UK, where he had worked on a cyclotron with similar energy levels. Palmieri was a post-graduate student, registered for his PhD in physics at Harvard. The focus of their joint research programme was to be the scattering of polarised protons but, before their experiments could commence, they had to build some specialised equipment. This enabled Allan to concentrate on other projects and activities during the Fall term.

Allan had time to attend an illuminating set of lectures on scattering theory by Roy Glauber who, at the age of just 18, had worked with Robert Oppenheimer, John von Neumann and Richard Feynman on the Manhattan Project at Los Alamos. Glauber followed a tradition established by his own thesis supervisor Julian Schwinger (one of Barbara's lecturers when she had studied physics at Radcliffe ten years previously), of lecturing late in the morning and then repairing with a few students to Chez Dreyfus, where an excellent lunch could be purchased for just 99 cents.[6] Among the post-graduate students who enjoyed this convivial arrangement was Shelly Glashow, who was to appear on the stage with Allan in Stockholm over two decades later. Glauber himself would have to wait considerably longer, until December 2005, before he won the Nobel Prize for Physics for "his contribution to the quantum theory of optical coherence".[7]

One piece of theoretical research that Allan had initiated in Cape Town, and was able to complete during his first few months at Harvard, was a systematic derivation of the mathematics pertinent to the calculation of Fourier transforms in cylindrical coordinates. He recognised that following the ground-breaking publication on the structure of DNA by James Watson and Francis Crick,[8] there was considerable interest in the helical structures found in certain molecules of biological origin. Allan drew attention to the use of the delta-function, introduced by his old Cambridge University lecturer Paul Dirac in 1930, for the description of curves and surfaces in three-dimensional space. Prior to submitting the manuscript for publication he shared an early draft with an old colleague from the Cavendish Laboratory, Dr Bill Cochran, who provided constructive criticism. The paper was subsequently published in *Acta Crystallographica* the following year,[9] and was the first of Allan's peer-reviewed articles to demonstrate his undoubted skills as an accomplished mathematician.

Allan also had time to turn his attention to the problem he had identified when working with Dr Grieve earlier in the year. He realised that for at least sixty years scientists had known that when X-rays pass through a homogeneous material, the intensity of the rays is attenuated exponentially. He also assumed, incorrectly as it turned out, that this finding had been generalised to inhomogeneous materials. A search of the relevant literature did not yield any joy. So he started to think about the problem from first principles.[10] He imagined shining a thin beam of X-rays through a cross-section of a patient's head (see Figure B.1). Allan knew it was possible to measure the intensity of the beam before it entered the head and also after it had been attenuated by the brain tissue and passed through on the opposite side. The key was then to be able to calculate the amount of attenuation (or absorption) at each point in the brain along the line. This became known as Allan's 'line integral problem', the solution of which would occupy him, on and off, for the next fifteen years.

Figure 5.1   The Harvard University Cyclotron with Lee Davenport (left) and Norman Ramsey (right), 1949.

By early 1957 the specialised equipment for the polarised proton experiment had been built and their joint research project on the cyclotron could begin. Their principal experiments were designed to determine the polarisation in proton-proton scattering as a function of laboratory angles (from 10° to 45°) and for various energy levels (from 90 MeV to 146 MeV).[11] To the chagrin of his postgraduate students who were working on molecular beam projects, Norman Ramsey took his turn to run an eight-hour shift on the cyclotron, sharing duties with Dick Wilson, Joe Palmieri and Allan. Ramsey's only stipulation was that his shift should be timed to coincide with the first bus between his home and Harvard.[12]

One Saturday evening Allan was doing the night shift, having started at midnight. Wilson came in at 8:00 am on the Sunday morning to relieve him and discovered that the cyclotron was not working.[13] Allan had lost the beam. Wilson established that the magnetic current had gone too high because a regulator was no longer functioning properly. Ramsey popped in to see how the experiment was running and he soon figured out what they needed to do to fix the problem. A chain link that was part of a mechanical feedback system had broken and Ramsey quickly redesigned the system: he replaced the link with an O-ring, and sent Allan off to the machine shop where he manufactured a pulley wheel on the lathe. By 10:00 am the cyclotron was again functioning properly and the team was back in action. After a brief consultation with his fellow physicists, Allan headed home for bed and a well-earned rest.

Allan held Norman Ramsey in the highest regard, not only for his scientific work but also for his service to physics and to the nation.[14] Ramsey had graduated from Columbia University in 1935 at the age of 20 with a degree in mathematics. Columbia then awarded him a fellowship to study at Cambridge University, where he enrolled for an undergraduate degree in physics, enabling him to rub shoulders with some of the great physicists at the day: JJ Thomson, Rutherford, Cockroft, Eddington and Dirac. He returned to Columbia where he collaborated with Isador Rabi for his PhD on the magnetic resonance method applied to molecular beams. During the war Ramsey worked on the development of radar and also spent time on the Manhattan Project at Los Alamos.

Shortly after the war ended, in 1947, he was recruited to Harvard University, where he established a molecular beam laboratory with the intent of doing accurate magnetic resonance experiments. However, he had difficulty in obtaining magnetic fields of the required uniformity. Ramsey was "inspired by this failure" and so he invented the separated oscillatory fields method, which had many applications, including the hydrogen maser and other atomic clocks.[15] Forty years later this work would be recognised with a Nobel Prize in physics, an award which many felt was long overdue.

During his time at Harvard, Allan was able to attend meetings of the American Physical Society in New York and Washington, DC, and he also delivered seminars on his polarisation work at institutions in the New England area.[16] His research on proton-proton scattering was subsequently published in the *Annals of Physics* and *The Physical Review*,[17] which provided one measure of the success of Allan's sabbatical leave. He was also still grappling with his line integral problem, an activity which did not escape Norman Ramsey's notice:

> Allan Cormack was a research fellow, working with Dick Wilson and myself. We were paying him from our funds. And, we knew he was boondoggling some of the evenings. He was doing calculations on his own. But that was okay. We let him do it. Well, it turned out that what he was doing was working on the theory for interpreting CAT scans![18]

Allan's ability was also recognised by Julian Knipp, head of the newly reconstituted physics department at Tufts University, who offered him a position as an assistant professor. The university had recently celebrated its centenary, having been established in 1852 on a grassy hill linking two suburbs, Medford and Somerville, a few miles north of downtown Boston. Its first President had been Hosea Ballou, a Universalist clergyman and historian, who was committed to scholarship. Allan was immediately impressed with Tufts but still needed to think about the offer.

There were four factors that Allan considered when deciding whether to accept the position.[19] First, it would enable him to continue working at the Harvard Cyclotron Laboratory, just three miles away from the Tufts campus which was located up the road from Cambridge in Medford, Massachusetts. Second, it would give him ready contact with some of the great centres of science instead of being secluded in a small enclave on the southern tip of Africa. Third, the move would provide regular contact with Barbara's family. Finally, it would enable the Cormacks to bring up their children in an environment different from that which prevailed in South Africa, a country which Allan thought was doomed to trouble. There was really no choice but to accept the position at Tufts although this did not prevent some pangs of conscience on leaving his native land at the age of 32.

On 23 June 1957 Allan wrote a letter to Mr JG Benfield, registrar of UCT, tendering his resignation as a lecturer in physics.[20] He also communicated with Walter Schaffer, who by now had become head of the physics department, outlining plans which he hoped would not disrupt the teaching in the department too much. Professor RW James, as Acting Principal, wrote to Allan on 4 July, saying he was sorry the Cormacks had decided to stay in the United States, but hoping the decision was the best one, "as indeed it may well be".[21] James also highlighted two potential problems for Allan. First, he quoted the rules governing resignation – which required six months' notice – and suggested the possibility that Allan may have to refund the whole or part of the salary he had received while on study leave. A decision on this issue would be made by the Senate (all the professors) and Council (the governing body) of UCT. Second, James reminded his young colleague that one of the conditions of the CSIR fellowship was that Allan should return to South Africa after his period overseas.

In early July Allan flew back to South Africa via England, leaving his family behind in Boston. In England he spent a few days at Harwell and at Cambridge. Much of his time at Harwell was spent with Godfrey Stafford and the cyclotron group, discussing matters of common interest since both laboratories (Harvard and Harwell) had machines of the same energy level.[22] Allan then flew home to Cape Town to await the Council's decision. Fortunately, a compromise was reached. Instead of six months' notice, the Council was prepared to accept three months; furthermore, they did not require Allan to repay any of his salary. This now meant that he would be able to return to Boston at the beginning of October, in time to take up his position at Tufts shortly after the beginning of the Fall term. The three months in Cape Town would provide Allan with an ideal opportunity to pursue the solution to his line integral problem.

While he was 'boondoggling' at Harvard, Allan had managed to solve the first part of his problem: recovering the attenuation coefficients for an object with circular symmetry. Having developed the initial theory, he was eager to test his ideas experimentally. After explaining his plans to his fellow lecturer Robin Cherry, Allan visited the mechanical workshop to discuss his requirements with the technicians, with whom he was very popular.[23] The workshop staff included its head, George Laing, as well as Jack Heald and Chick Fowle. Allan shared with them a sketch of his circularly symmetric absorber or 'phantom', a uniform disc of aluminium (representing brain tissue) surrounded by a wooden annulus (representing the bony skull). Instead of X-rays, Allan used gamma-rays, the source being Cobalt-60 in the form of a small silver-plated rod, one centimetre long and a millimetre in diameter. A pair of lead collimators was used to 'guide' the gamma-rays through the phantom, and a Geiger counter recorded the raw data (Figure 5.2).

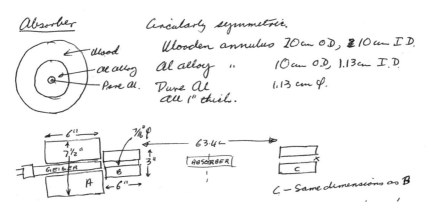

Figure 5.2   Allan Cormack's sketches of a wood and aluminium absorber (top), and a Geiger counter plus lead collimators (bottom), 1957.

Allan gathered his first set of data in the basement of the physics department at UCT on Tuesday 3 September 1957 and by Friday 6 September his experiment was complete.[24] When he compared the experimental data with his theoretical predictions, he was delighted to discover the agreement was very good (see Figure B.2 on page 226). However, there was one small puzzle. There appeared to be an anomaly in the data near the centre of the aluminium disc.[25] He questioned Laing and the other machinists, who revealed that they had inserted an aluminium peg of slightly different density (one or two percent lower) at the centre (Figure 5.2). This serendipitous finding was particularly encouraging for it meant that Allan's technique could detect small density differences, a crucial requirement for the diagnosis of brain tumours and other lesions.

While he was busy with his experiments in Cape Town, Allan continued to grapple with his line integral problem, working on the solution to the general case where the phantom lacked circular symmetry. Thinking that such a 'natural' problem must have been solved by a 19th century mathematician such as Cauchy or Riemann, he asked around at UCT.[26] One of the mathematicians he consulted was an old climbing companion, Denis Rutovitz, who had been an undergraduate at Cape Town in the late 1940s but left to spend time as a pilot in the Israeli Air Force. He returned to UCT in 1956 to pursue a doctorate in mathematics and, although he was unable to answer any of Allan's questions, he became most enthusiastic about the project. Rutovitz was married to a medical doctor, Eugenia Cheesemond, and the two of them tried to drum up support among her radiology colleagues at Groote Schuur Hospital about the potential of Allan's discovery, but to no avail.[27]

Allan also discussed his project with Donald B Sears, professor of mathematics at UCT, and left a copy of a draft paper with him. Sears responded on 24 October 1957 – by which time Allan had already started his job at Tufts – verifying the correctness of the results, and saying that he had found a valid solution for the general case.[28] By this stage Allan had found such a solution himself and, with the demands of his new position and trying to catch up with what had happened during his three month absence from the Harvard Cyclotron Laboratory, he did not follow up on Sears' note. Sears, however, had suggested earlier to Allan that he write to Professor EC Titchmarsh, a distinguished mathematician at Oxford University. Eighteen months were to elapse before Allan followed up on this suggestion. In mid-1959 he was invited to attend a conference at University College London on 'The Few-Nucleon Problem' and, since he planned to spend a few days with Godfrey Stafford and his wife Goldie near Oxford, he thought this would be a good time to speak to Titchmarsh. Allan sent him a draft manuscript with the comment:

> I would like to publish the work but I have a great fear that the problem has either been solved and is unknown to physicists, or that it has not got a solution, even though I appear to have found one. I have shown it to a couple of mathematicians and theoretical physicists who have said that it "seems to be all right" but I would prefer to have the opinion of someone of your erudition before risking making a fool of myself in print.[29]

Titchmarsh sent a cordial reply, saying that as far as he knew Allan's work on the absorption coefficients was novel, and invited him to visit.[30] On 13 July the two of them got together. Titchmarsh said he believed the solutions to be correct and reassured Allan with the comment that "there are many more problems in mathematics than there are mathematicians to solve them".[31] During this trip Allan also met up with Rutovitz and Cheesemond, now living in England, who told him that they had been trying to persuade influential British radiologists of the clinical potential of Cormack's method. Once again, nobody was interested.[32]

Also recruited to the Tufts physics department by Julian Knipp in the summer of 1957 was Brenton Stearns, who had just completed his PhD at Washington University in St Louis, Missouri.[33] He and Allan soon became firm friends when they encountered one another in the Fall after Allan arrived from Cape Town to take up his new position. Brent had been involved in the development of the Cambridge Electron Accelerator, a joint project between Harvard and the Massachusetts Institute of Technology, while Allan continued his research at the cyclotron. The two young assistant professors decided to collaborate in building instrumentation for experiments to be done on the new accelerator. Their first project was a Čerenkov counter, an instrument based on the principle that a charged particle moving through a transparent material at the speed of light generates a cone of light that is aligned with the particle's path. To produce this effect they needed to find a source of lead glass with a high atomic number and a relatively high index of refraction.

Allan suggested that they visit Cornell University in upstate New York, where the physicists had built small counters for work on their lower energy accelerator. What the Tufts men learned when they got to Cornell was not encouraging: small lead inclusions in the glass decreased the efficiency of the counter, while the manufacturer, the renowned Corning Glass Works company that was located just up the road, said this was the best they could do. Brent and Allan flew home to Boston empty-handed but undeterred. They had heard that it might be possible to grow transparent crystalline masses of lead sulphide, and they set up a graduate student in a drafty attic in Robinson Hall, the home of the physics department, to begin the process.

Although Allan and Brent had some limited success in creating a transparent mass with dark inclusions, one of their physics colleagues, Dr Kathryn McCarthy, got wind of their project. Her father was an inspector for the Massachusetts health department and he informed them that they were doing something rather hazardous, and should stop immediately, which they did. It would be another decade, while on sabbatical at the Lawrence Radiation Laboratory in Berkeley, California, before Allan had an opportunity to work on the design of an improved Čerenkov counter.[34]

One of the physicists that Allan met at the Harvard Cyclotron Laboratory during his sabbatical year was Andreas Koehler. Their acquaintance led to a life-long friendship, a special relationship of mutual respect that would see them collaborate on a variety of projects over the ensuing 40 years. Andy became a sounding board for many of Allan's ideas, which were invariably shared during their lunch breaks at the cyclotron:

> What I got from him was the encouragement to persevere, to do things that I wasn't trained to do. I think what Allan may have gotten from me was the enthusiasm of the novice. I was always thrilled if he would take the time to explain to me what he was doing and I think he enjoyed that. At some point, he'd whip out a piece of paper and draw me a diagram and I'd do my best to understand it. Even if I didn't understand it, we still communicated.[35]

One of the people with whom Allan and Andy collaborated at the cyclotron was Mildred Widgoff. Mildred had experience with nuclear emulsions for recording the tracks of particles and although the technique was out of favour by then, having been replaced by modern instrumentation, she persuaded her colleagues that it still had some real value. They set up an experiment measuring the scattering of protons off deuterium and by understanding the scattering process, they learned about neutron-neutron interaction. By studying the 'spectator particles' – the protons were knocked out but did not get involved in the reaction itself – they were able to learn something directly. All during this period, Allan continued to plug away at his line integral problem. Years later, Widgoff recalled:

Allan, Andy Koehler and I collaborated on an experiment at the Harvard Cyclotron at about the time I moved from Harvard to Brown in the early 1960s. It was a most pleasant and satisfying joint effort. It is one of my most amusing memories, Allan telling Andy and me over dinner breaks about this little calculation he was doing for the fun of it. It seemed such a harmless hobby![36]

It was at this time that Andy embarked on a programme of research that would redefine the *raison d'etre* and significantly extend the life of the Harvard Cyclotron Laboratory. The inaugural director, and the person responsible for the main design features of the cyclotron, was Robert R Wilson, known as Bob. In a landmark paper, published in 1946, Wilson pointed out that a high energy proton beam would be effective for the treatment of cancerous tumours, by producing small regions of high ionisation in the tissue.[37]

This phenomenon – known as the Bragg Peak in honour of its discoverer William Bragg – meant that the proton beam produced a greater ionisation density at the end of its range, typically at a penetration of about 12 centimetres, at which distance the energy suddenly dropped to zero. By arranging for the tumour to be bombarded from different directions, while ensuring that the peak always coincided with the lesion, maximum damage could be inflicted on the tumour while the damage to the surrounding healthy tissue would be minimised. Wilson, like many of the brightest physicists at the time, had been seconded to the Manhattan Project in New Mexico during the war and stated that he had been motivated to devote some time to this medical application as "atonement for my involvement in the development of the bomb at Los Alamos".[38]

It would be fifteen years before this idea was resuscitated by William 'Bill' Sweet, head of neurosurgery at the Massachusetts General Hospital, and his colleague Ray Kjellberg. They approached the director of the Harvard Cyclotron Laboratory, William 'Bill' Preston, who in turn asked Koehler to explore the feasibility of using the cyclotron for cancer treatment. Andy set about conducting a series of experiments, including extensive use of monkeys, in order to understand the ionising effects of the proton beam on brain tissue. He designed a stereotactic frame for positioning the patient so that the centre of the tumour would always be located at the Bragg Peak and he developed a telescoping water absorber to compensate for any change in the amount of tissue the protons had to penetrate to reach their target. The first patient, a young girl with a large pituitary tumour, was treated in 1961, and the team's early successes soon led to the regular use of proton beams for the treatment of intracranial lesions.[39]

Satisfied that he had solved the theory of his line integral problem, Allan turned his attention to testing his ideas experimentally. In Cape Town he had gathered data on a phantom with circular symmetry (Figure 5.2), but a far more interesting problem was the asymmetrical case. He approached George Wallace, a short, wiry man with a typical Scottish brogue, who had worked in the machine shop at the Tufts physics department for many years.[40] Allan shared with Wallace his simple sketches of asymmetric phantoms. Each was a disk with a diameter of 200 millimetres, manufactured from Lucite, in which were embedded smaller metal disks of iron and aluminium (Figure 5.3).

Based on Cormack's design, Wallace also built a simple instrument – in effect, a prototype computerised axial tomography, or CAT, scanner – that would enable Allan to gather data for his asymmetrical phantoms. There were two key features to the design: first, it allowed the phantom to be translated horizontally through a beam of gamma rays; and second, the phantom could be rotated about a vertical axis (Figure 5.4). Since it was only necessary to gather data for half the width of the phantom and half a revolution, measurements were to be made at 41 horizontal positions (from 0 to 100 mm in 2.5 mm increments) and at 25 angular positions (from 0 to 180 degrees in 7.5 degree increments). Before Allan began experiments to gather a set of 1025 data points with his new apparatus and phantoms, however, he finally decided it was time to test the waters with his earlier work on the phantom with circular symmetry.

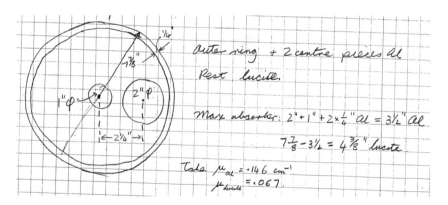

Figure 5.3   Allan Cormack's sketch of a phantom made from Lucite and aluminium, 1963.

On 22 January 1963, more than five years after he had originally gathered the data in Cape Town and solved his line integral problem for the case of circular symmetry, Allan submitted a manuscript to the *Journal of Applied Physics*, entitled 'Representation of a function by its line integrals, with some radiological applications'. After a few minor changes suggested by the editor and referees, the paper was published later that same year, in September.[41] At the end of the article Allan acknowledged the contributions of two people: Muir Grieve, who had identified the necessity for a solution to the absorption problem; and Andy Koehler, who had pointed out that the solution could be used in the coincidence problem.

In the coincidence problem, a patient was given a positron emitter, typically copper-64, which concentrated on abnormal brain tissue. The positrons were then annihilated and pairs of photons were emitted in opposite directions, while two counters, located on opposite sides of the head, measured the coincidence rate and therefore the rate of annihilation along the line joining the counters. This technique was the forerunner to positron emission tomography, or PET, scanner,[41] and had been pioneered in the early 1950s by Bill Sweet and Gordon Brownell, the latter a medical physicist from MIT who was affiliated with the Massachusetts General Hospital.[43] Allan also highlighted a third application of his solution, determining the ionisation distribution when using protons and the Bragg Peak to treat tumours, which he had also learned about through his collaboration with Koehler.

By the Spring of 1963 Allan was ready to begin the serious work of gathering the first measurements with his asymmetric phantom and turntable scanner (see Figures 5.3 and 5.4 and Figures B.3 and B.4). Assisting him were two of his students: David Hennage, an undergraduate majoring in physics; and then later Bob Schneider, a postgraduate and Allan's first PhD student, who worked with neutrons at the cyclotron.[44] The first set of data was gathered in a building adjacent to the Harvard Cyclotron Laboratory using a cobolt-60 source, all arranged by Andy Koehler. As Cormack later recalled:

> Andy believed, I think correctly, that he would be violating the Atomic Energy Commission's regulations by letting me (who did not have an AEC licence for such a source) transport the source to Tufts and use it there. He was prepared to let me use it in premises under his control provided that the less that was said about it, the better![45]

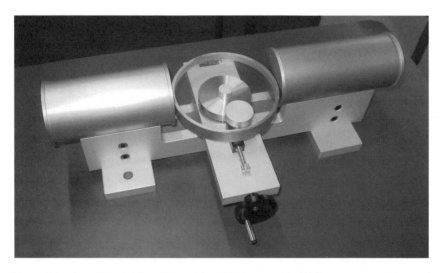

Figure 5.4   A replica of Allan Cormack's original turntable CAT scanner manufactured by the author, 2005. The original is in the Smithsonian Museum, Washington, DC.

Allan dropped Hennage off at the cyclotron facility in the early evening and returned later that night to fetch him. Unfortunately, Hennage had rotated the phantom symmetrically around the offset 'tumour', but had forgotten to move it horizontally, and so they did not have a complete set of data.[46] They therefore had to return a second night to complete the experiment. The next step was to reduce the data using Allan's algorithm based on line integrals, and to combine the experimental results with theoretical predictions. Hennage took a four-day course on FORTRAN programming and then began to write the computer code that would implement Allan's equations. He initially punched out the cards at Tufts before cycling over to the MIT Computation Center where the program and data were submitted. Since Hennage was new to programming, he needed to perform multiple submissions before the program compiled successfully and ran. The results, however, were not quite right. Hennage then reprogrammed the algorithm using double precision mathematics and finally the output showed a good match between the known attenuation coefficients for aluminium, Lucite and air, and those predicted by data plus theory (see Figure B.4 on page 228). Allan was delighted.

On the afternoon of 22 November 1963 Hennage was walking from the physics building up to the main Tufts campus when he met some postgraduate students who told him that President Kennedy had been shot. David immediately rushed up to Allan's office to share with his professor the news that "The President has been shot!" A puzzled look came across Allan's face and he asked the young undergraduate, "Why on earth would anyone want to shoot President Wessell?" Slowly it dawned on him that it wasn't the President of Tufts, Nils Yngve Wessell, who had been shot, but the President of the United States! A little later that same afternoon Allan presented a seminar to the staff and students in the physics department, during which he spoke about the mathematical solution to the problem he had been working on for over seven years. Bob Schneider, Allan's PhD student, recalled that his hand was shaking so much he could barely take notes but, undeterred, Allan lectured calmly on![47]

Having suggested that Allan's solution to the line integral problem would apply to PET scanning, Andy Koehler urged him to meet with Gordon Brownell and his colleagues at the Massachusetts General Hospital. The plan was to try and persuade them that there were better ways of taking and interpreting their data. So he met with Brownell and Saul Aronow at 10:30 am on Tuesday 17 December 1963 and eagerly shared with them his results, which had not yet been published.[48] Allan discovered they were using a coincidence circuit with a temporal resolution of several microseconds whereas he and Koehler were using a circuit with a resolution in the nanosecond range. Unfortunately, the two medical physicists showed no interest in any of Allan's work or ideas and he left the meeting distressed, upset he had been unable to convince them that with a little extra effort they could have significantly improved their PET scanner's ability to locate tumours. He did, however, retain a record of his appointment, a photocopy of his diary for that week.[49] Many years later Allan met Aronow at a meeting and Aronow acknowledged that for some time he had been kicking himself for having ignored the advice, freely given at their earlier meeting.

In early 1964, Andy Koehler and the two Bills – Sweet and Preston – were busy preparing a grant application for submission to the National Aeronautics and Space Administration (NASA), to support their research on proton therapy at the cyclotron. The peak of the energy spectrum for cosmic ray particles was 150 MeV and NASA did not yet understand the effect that 150 MeV protons would have on their astronauts.[50] Allan was invited to submit a brief proposal on his own research which he did, including the unpublished data that he had gathered and analysed with David Hennage the previous summer.[51] The focus of his project was to be an adaptation of his line integral problem to PET scanning. He planned to investigate the feasibility of his approach, but not necessarily to bring it to a state of clinical usefulness, and requested a modest budget of $6,600. At the time NASA was "just gushing with money" and the entire proposal was funded except for Allan's small part![52]

By April 1964 Allan was ready to submit a second paper to the *Journal of Applied Physics*. Designated as part II of his first paper, it shared the same title. The original work had been extended in several ways: the entire plane, rather than a simple unit circle, was described; two separate mathematical solutions to the line integral problem were given, including useful computational methods; and results were presented for a phantom that lacked circular symmetry (Figure B.4). Allan Cormack's most significant paper was published in October that year; the succinct abstract was a model of lucidity:

> A method is described for determining a variable gamma-ray absorption coefficient in a sample from measurements made outside the sample, and the applicability of the method to other radiological problems (*e.g.*, positron scanning) is pointed out. An experimental test of the method is described, and it is concluded that the method works well.[53]

When Allan returned from Cape Town in the autumn of 1957 the Cormack family had moved into rented accommodation, a house at 1 Woodside Road in Winchester, a middle-class suburb just a few miles north-west of the Tufts campus. They were soon ensconced in the comfortable suburban lifestyle, enrolling Margaret in the nearby Wyman Elementary School, and acquiring, in quick succession, three cats – Spotty, Ginger and Midnight, or Middy for short – all named for their physical appearance.[54] Within two years they had moved again, to another rented house in Winchester, the lower floor of a large three-story building at 45 Fletcher Street. Their next acquisition was Charlie, a large pure-bred poodle, who was given to them by Dr Asger Aaboe, one of Allan's colleagues, who was moving to Yale. Charlie was an inveterate chaser of cars and cats, and liked to wander, but would always return home if one of the family shook the box of dog biscuits loudly enough. Allan purchased a pair of clippers and clipped the dog himself but it was never the traditional 'poodle cut'. The result was that Charlie seldom looked like a poodle. When a visitor or passerby enquired what kind of dog he was, Allan would make up the name of an imaginary breed. The reply was invariably "Oh yes, of course", which caused much mirth in the Cormack household.

The Seavey family was originally from New Hampshire and in the 1930s

Warren Seavey had purchased a farm in Alton, near his family's former home, on the shore of Lake Winnipesaukee.[55] True to its name, which translates as 'deep water in a high place', the lake stands in the shadow of the granite edifice of Mount Major.[56] The farm was a special retreat for the extended Seavey family, serving as a base both in the summer and in the winter, when some of them would go skiing in the nearby mountains.[57] Warren and Stella Seavey spent much of their retirement up there, staying in an outside cottage which had been the old blacksmith shop and to which they had added a kitchen and bathroom.

The Cormack and Seavey families all stayed in the main house, eating their meals together at a long dining table, sharing the simple pleasures of wild blueberries and corn on the cob.[58] Twice a day they swam in Lake Winnipesaukee. A rowboat that they launched at Roberts Cove allowed them to enjoy picnics on a nearby island. A favourite annual outing was to Devil's Den, a natural cave cut into the granite rock, where, after a strenuous climb, they could eat their picnic lunch and enjoy the spectacular scenery. For Allan, Alton was an idyllic place, where he would work late into the evening and then rise late the following morning. It was a place for conversation, to enjoy the company of his family and to recharge his batteries during the long summer recess.

In 1960 Allan was promoted to associate professor and, with Barbara working part-time as a mathematical analyst at Harvard College Observatory,[59] they had accumulated sufficient capital to make a down payment on their first home, 18 Harrison Street, Winchester. It was located several blocks from their rented house in Fletcher Street, which made for an easy move. The spacious Victorian-style home had a large lawn for Charlie, a fishpond and a double garage for the family that would soon acquire its second car. This would be the Cormack family home for the next 38 years. It was just a mile from downtown Winchester, so Margaret, and later Jean, could walk to school. The parents encouraged both their daughters to learn to play a musical instrument and both of them chose the violin. Allan and Barbara were late adopters of television, only acquiring their first TV set in 1963.[60] Initially, there was a strict rule of only one hour's viewing per night, although this restriction excluded the evening news when the increasingly popular anchorman, Walter Cronkite, would inform the American public of the day's events.

The lack of a TV meant that the Cormacks missed the first-ever televised presidential debate between Senator John Kennedy of Massachusetts, the Democratic nominee, and the Republican nominee, Vice President Richard

Nixon, on 26 September 1960. They also did not see the report on the Soviet cosmonaut Yuri Gagarin, who became the first human to travel into space in Vostok I on 12 April 1961. Although the USA responded by sending astronaut Alan Shepard into space less than a month later, Gagarin's flight had dealt a serious blow to the young President Kennedy. Shortly thereafter, before a joint session of Congress, he boldly declared "I believe this nation should commit itself to achieving the goal, before this decade is out, of landing a man on the moon and returning him safely to Earth".[61]

Kennedy's challenge would galvanise the American people who rallied their support for NASA and its space exploration endeavours. It also led to a boost for the field of physics with the concomitant expansion of academic programmes at Tufts and elsewhere. John Glenn became the first American to orbit the Earth on 12 February 1962, marvelling at the beauty of the sunrise before sweating through a fiery re-entry into the Earth's atmosphere. Twenty years later, when Glenn was a US Senator, he and Allan would have an informed exchange of ideas regarding the role of the Federal government in the support of scientific and technological research.[62]

Despite her advancing age, Allan's mother was in good health and celebrated her 80th birthday in December 1963 (Figure 1.2). She was a regular correspondent, keeping her son informed about developments in Cape Town and elsewhere in South Africa, and filling in details on their Scottish ancestors. She especially enjoyed receiving letters from her two American granddaughters, learning about their various school projects and other social activities, in particular their music.

Barbara was enjoying her work at the Harvard Observatory where she was involved in the measurement and reduction of photographic meteor trails to determine the spatial trajectories and velocities of meteors. This gave her an opportunity to learn FORTRAN programming to reduce the data although she did not apparently assist Allan with his research. The job came to an end when Barbara gave birth to their son Robert on 8 August 1964 (Figure 1.2). This was an auspicious year for Allan as well. Not only did he publish the definitive paper on his line integral problem, but he was also promoted, at the age of 40, to full professor of physics at Tufts University, and elected a Fellow of the American Physical Society.

Chapter 6

Finding Radon and his Transform

Browsing in the mathematics library in 1971, I came across Fritz John's little book on plane waves. There on page nine was 'my' line integral problem with reference to a man called Radon, writing in the journal of the Royal Saxon Academy of Sciences in 1917.

Allan Cormack[1]

In the mid-1960s, prior to the widespread availability of photocopying machines and well before the advent of the internet, the standard practice for academics wanting a copy of a published article was to request a reprint from the author. The number of reprint requests, normally in the form of a simple post card, was a crude measure of the impact that the research had made on its field. For Allan's two papers in the *Journal of Applied Physics* he received a total of just three requests.

One of these requests came from Professor Claude Jaccard of the Université de Neuchatel in Switzerland.[2] Jaccard was the director of the Swiss Federal Institute for Snow and Avalanche Research, based in Davos. Allan was amused to think that his line integral technique might find application in the search for a skier buried in the snow. The only problem was how to get either the detector or the source into the mountain and under the snow![3]

Although this story would always elicit a good laugh, Allan believed that it served another purpose, namely to point out that articles intended for one particular audience could be ignored by that audience but discovered by someone in another field who was alert to possibilities. While radiologists might have missed the significance of his work, since they would not normally have read the *Journal of Applied Physics*, a physicist involved in snow research recognised the interconnectivity of the scientific disciplines.[4] In fact, Jaccard was interested in Allan's analysis because it enabled him to use a microscope to study the polycrystalline structure of snow and ice.[5] The structure influenced the material properties of snow, which would, in turn, determine its behaviour when subjected to the mechanical stresses generated during an avalanche.

Allan had waited several years, from his first experiments at the University of Cape Town (UCT) in 1957 to his research at Tufts in 1963, before publishing his algorithm and findings. If he had planned to develop his ideas for profit, he would have had to lodge a patent application with the United States Patent and Trademark Office *before* the publication of his work in journal form. When later asked why he had not followed this course of action, Allan responded:

> Well, it never occurred to me. You know, I went into physics to do physics, and not to make money or that sort of thing.[6]

As an undergraduate at UCT in 1943 Allan had switched majors, from electrical engineering to physics, as his interest and focus changed from applied to basic science. He was far more interested in solving a difficult scientific problem – such as his line integral problem – than turning the solution into a practical product. With the lack of interest shown by the radiological community in his work, his research focus returned to particle physics.

Allan was collaborating with his PhD student, Robert Schneider, using the neutron beam of the Harvard Cyclotron Laboratory (HCL) to measure the cross-sections of the elements aluminium, copper, lead and uranium.[7] He was also continuing his research with Andy Koehler and Mildred Widgoff, the latter of whom was spending a sabbatical at CERN in Switzerland. In the adjacent building to the HCL was the Cambridge Electron Accelerator (CEA), jointly operated by Harvard University and the Massachusetts Institute of Technology (MIT), and funded by the Atomic Energy Commission to study the physics of high energy interactions of electrons and photons. In the spring of 1965 there was a fire at the CEA, and spreading oil ignited a condenser which, located near the linear accelerator, exploded.[8] Fortunately the damage was relatively minor, but the incident presaged a far more devastating event later in the summer.

On 4 July 1965, the majority of Americans were celebrating Independence Day while Allan and his colleagues Andy Koehler and Bernard Gottschalk were gathering data for one of their experiments. They departed the HCL at about 7:00 pm, leaving a student, Val Kursis, to acquire data during his night shift.[9] Next door in the CEA a technical team from MIT was in the experimental hall monitoring a programme for filling the Bubble Chamber with liquid hydrogen. The chamber was filled to about 95 percent of its 500 litre capacity.[10] The cryogenic system used low temperature helium gas, circulating through condensers in the upper portion of the chamber, to liquefy the hydrogen gas. Without warning, at 3:32 am on 5 July, the Bubble Chamber exploded, lifting the CEA's roof off its supports.[11] The reinforced concrete panels, topped with a mixture of tar and gravel for insulation, broke and shattered, the debris falling onto the scientists and their equipment in the hall below.

Fred Gale, a cryogenics technician, was hurled by the force of the blast through a doorway. Disregarding his own safety, he rushed back to the scene of the explosion and fire where he turned off the helium compressor and then led three of the injured scientists to safety. Within minutes the Cambridge fire brigade was on the scene. A postgraduate student, woken by the explosion, raced over to the CEA where he encountered some hostile firemen who were trying to break down a door to gain access to the building. Their hostility soon turned to gratitude when he explained that he had a key that would open the door![12]

Next door in the HCL, Kursis telephoned Gottschalk to announce that the CEA had blown up and the cyclotron had lost its beam.[13] A quick check on the control panel showed that a safety switch was open and a tour of the facility revealed that the back door to the laboratory had been blown open by the blast from the CEA. With the door shut, Kursis soon had the beam restored and resumed his experiment. However, his data gathering was short-lived. He discovered that there were firemen clambering on the HCL roof and, since this was a high radiation area, he turned off the cyclotron.

Many of the HCL staff, including Allan and Koehler, had by now been notified of the disaster and drove down to see what was happening.[14] An impromptu staff meeting was held, the main agenda item being a discussion of what impact the explosion and fire might have on the activities of the CEA and HCL. Eight people had been injured, three of them seriously. For his act of heroism, Gale would later receive a Citation from the Atomic Energy Commission.[15] Tragically, one of the scientists died two weeks after the incident. It was a sobering reminder to Allan and his colleagues that research in nuclear physics carried inherent dangers. The damage to the

CEA was established to be $1 million and it would be almost a year before the laboratory was again operational. Once power had been restored to the area two days later, the HCL was up and running and it would continue as a functional research and clinical facility for a further 37 years.

By the middle of the following year, the summer of 1966, Allan and his family had been living in the USA for ten years. He decided that it was time to apply for American citizenship. As part of his preparation for this process, he borrowed his daughter Margaret's social studies textbook.[16] Having studied, among many topics, the constitution of the USA, the questions he was asked were extremely straightforward: "Who is the President of the United States?" and "Who is his Vice President?" Allan had no difficulty with the answers – Lyndon Johnson and Hubert Humphrey. At his final immigration hearing in downtown Boston he was accompanied by Barbara and two of his colleagues from the physics department at Tufts, Kathryn McCarthy and Brenton Stearns.[17] Although Allan relinquished his South African passport, he never severed his ties with the country of his birth nor did he ever cease thinking about his Scottish heritage.

Stearns had spent the academic year of 1964–65 on sabbatical leave at the Lawrence Radiation Laboratory, University of California at Berkeley. On his recommendation, Allan wrote to the co-directors of one of the research groups, Burton Moyer and Carl Helmholz, asking if he could spend his sabbatical at the lab in 1966–67. An invitation was extended and in August 1966 the family boarded a commercial flight at Boston's Logan airport and flew to Denver, Colorado. For the two younger children – Jean aged 10 and Robert, aged 2 – this was their first trip in an aeroplane.[18] Allan had arranged for the family car to be driven from Massachusetts to Colorado and so it was waiting for them at the airport.

From Denver they drove north to the Black Hills of South Dakota, then travelled due west, through Wyoming, to Yellowstone National Park and then on to the majestic mountains of the Grand Tetons. The Cormack family vacation, and their great trek across the Rocky Mountains, ended in California, at Yosemite. From there it was just a few hours' drive to Berkeley where they rented a house for a year from an academic who was on sabbatical in Germany.[19]

The Lawrence Radiation Laboratory was founded by Ernest Lawrence in 1931 and nine years later moved to its location on the hill above the Berkeley campus, with a commanding view of San Francisco and the Golden Gate Bridge.[20] It was here, on the 200-acre campus, that Lawrence built his 184-inch cyclotron, which was used during World War II to develop the principle behind the electromagnetic enrichment of uranium. Allan joined the Moyer-Helmholz group where his primary collaborator was a PhD student, David Ayres.

Ayres had recently returned from a two-year stint in Nigeria where he taught physics to undergraduates at Enugu. Like many other young Americans in the early 1960s, he had been inspired by John F Kennedy to join the Peace Corps. Ayres and his wife – also a Peace Corps volunteer – had returned from Nigeria to the USA via South Africa, so they identified with Allan and had much to talk about. Over 40 years later, Ayres recalled one of the anecdotes:

> I remember Allan's story of how his hospital job in Cape Town required him to meet incoming aircraft and retrieve medical radio-isotopes that were attached to their wing tips to minimise radiation exposure of the passengers. I never doubted that story, but as I write it down now, it does seem rather improbable![21]

Dave Ayres had completed a master's degree at Berkeley before his two-year sojourn in Africa and was now busy with an experiment to measure the lifetimes of positive and negative pi mesons (π^+ and π^-). A fundamental law of symmetry – the inversion of charge, parity and time or CPT – required that the lifetimes of π^+ and π^- should be equal. His earliest results suggested that the lifetimes were slightly different, which, if true, would have meant he had discovered a "discrepancy in physics even greater than the violation of parity conservation with which Yang and Lee had startled the world".[22] It turned out the experiment was fundamentally flawed and Allan had an opportunity to participate in a research programme to set the record straight.

A beam of pions (either positive or negative, depending on the polarity of the beam line magnets) was created by smashing a stream of protons from the 184-inch cyclotron into a beryllium target. The pions were focused into an elongated beam and, as they travelled along their 9-metre path, they decayed into muons. The flux of pions was measured at different distances along the beam line by means of a Čerenkov counter (Figure 6.1). The critical mistake made during the earlier experiment had been to fill the Čerenkov counter with ordinary hydrogen, where the interactions of π^+ and π^- were slightly different. When the hydrogen was replaced by liquid deuterium, the interactions were the same and the lifetimes were found to be equal, thus preserving symmetry. The precise values (approximately 26

Figure 6.1   Dave Ayres with the Čerenkov counter (left) and the 9-metre beam path where pions decayed into muons (right), 1966.

nanoseconds) were subsequently published in *Physical Review Letters*, with Allan as a co-author.[23]

While they were busy with these first experiments, Allan realised that the Čerenkov counter could also be used to measure the index of refraction of liquid deuterium. The team had to set the energy of the pion beam to just the right value to give the precise Čerenkov angle that would be collected by their optical system onto a large photomultiplier tube with high efficiency. Allan derived great satisfaction from working on an obscure topic such as this, and digging into it deeply enough to learn something interesting, rather than working on a large experiment to measure something that was considered fashionable and more visible.[24] He was delighted when the paper was later published in *Physica*,[25] a journal that he held in high regard. This was a productive sabbatical for Allan, in which he co-authored five papers with David Ayres and the group at Berkeley. They were to be his last publications in the field of particle physics.

The year in California enabled Allan and his family to take advantage of the state's great variety of places to visit. At Christmas time they travelled south, passing through the Redwood forests, the great trees dating back to before the time of Christ. In Los Angeles the children particularly enjoyed the day at Disneyland. Then, in the spring of 1967, the Cormacks ventured further afield, to Arizona, where their end destination was the Grand Canyon. Once they had taken in the awe-inspiring view of the gaping void, they discovered that it was possible to ride a mule part of the way down into the canyon (Figure 6.2). However, there was an age limit of 12. While Robert was clearly too young, Jean had just celebrated her 11th birthday a few months earlier. Margaret recalled her father lowering his voice and telling them, conspiratorially, that "This is a once in a lifetime opportunity so, just this one time, we'll say that Jeanie is 12!"[26]

Figure 6.2   The Grand Canyon: Jean and Margaret in front, Barbara at rear, spring of 1967.

Allan's sister Amy and her family – husband Len, daughter Jo and son George – had visited Boston earlier in the summer of 1967. Unfortunately the cousins did not get to meet one another although Allan did fly back briefly to see the Reads. The Cormacks returned to the East Coast at the end of the summer, having taken full advantage of the sabbatical year in California. Allan soon settled back into the routine at Tufts, teaching undergraduate physics and dutifully attending faculty meetings.

At home in Boston, around the dinner table, Allan shared his frustrations regarding the behaviour of some of his colleagues at Tufts, with a strict injunction to the family not to repeat anything he had said. When asked why he attended these meetings, he responded "If the sensible people stayed away, the idiots would have everything their own way!"[27] Allan was especially concerned about the health of the College of Arts and Sciences, in which the physics department was located and was vehemently opposed to the suggestion that their department should be transferred to the College of Engineering.[28] In March 1968 he was appointed chair of the physics department, a position he was to hold for the following eight years.

It had been ten years since Allan had last seen his mother. She was, however, a marvellous correspondent and wrote regular, lengthy letters to her younger son. Although by now in her early 80s, Amelia Cormack still had a wonderful memory, recalling details of the Millers, Murrays, MacLeods and Cormacks from the late 19th and early 20th centuries (Figure 1.2). She knew of Allan's abiding interest in his Scottish ancestry, and told him the story of the Tay Bridge disaster of 1879 and the involvement of her mother's uncle, Tom Miller, in this tragedy. She also reminded him of the occasion in 1938 when they were about to pay a visit to an old friend in Oban, only to discover that she had recently died, whereupon Allan had remarked, "Thank goodness, that's one you cannot visit!"[29]

Allan's older brother Bill was by now head of the department of electrical engineering at the University of the Witwatersrand in Johannesburg, while his sister Amy and her husband Len were still working for the University of Cape Town. In early 1968 Allan received the news that his mother had been diagnosed with breast cancer. Because the disease was already quite far advanced, the doctors decided that the tumours were inoperable. With her condition deteriorating rapidly, Amelia Cormack moved from her flat in Central Square, Pinelands, to her daughter's house in the same suburb, at 9 Margaret Avenue. The two Read children, 12-year-old Jo and 10-year-old George, shared a room so that their granny could have the comfort and privacy of her own bedroom.

Amelia MacLeod Cormack died on 15 November 1968, just five weeks short of her 85th birthday. Allan was unable to travel from Boston to Cape Town for his mother's funeral but he sat down and wrote a lengthy letter to his South African family. The passing of their mother, born on Scottish soil, marked the end of an era for the three middle-aged siblings.

During the first three years of his tenure as department chair, Allan was a diligent administrator (Figure 6.3). Although his research suffered – the only publications that came out were from his time at Berkeley – he took the responsibility of chairing his department seriously. For some time he had been concerned that the mathematics required to be taught to their postgraduate students stemmed from the period before the mid-1930s: Courant, Hilbert and the group theory of Wigner.[30] He started asking the mathematicians at Tufts what had been happening in the previous 30 to 40 years. He was given a number of references which he perused diligently but without reward. However, while browsing in the Tufts library in the summer of 1971, Allan stumbled across a book by Fritz John.[31] There on page nine he discovered 'his' line integral problem, with reference to the work of a man called Johann Radon.

Figure 6.3   Allan Cormack, the new department chair of physics at Tufts, 1968.

Radon's paper had been published in 1917 in German in the transactions of the Royal Saxon Academy of Sciences[32] and not, as Allan had previously speculated, in the late 19th century. Since the Tufts library did not have the transactions, he telephoned his elder daughter, who had just completed her freshman year at Radcliffe, the women's college affiliated with Harvard University. Margaret had started off majoring in physics – the same major as both her parents – with a view to specialising in astronomy and had passed the first year comfortably. However, she had soon realised this was not what she wanted to pursue and switched to linguistics. After receiving the phone call from her father, Margaret walked over to the Widener library at Harvard from Currier House and photocopied the article by Radon. Her interest in Germanic languages was to come in handy as she also translated the article for her father.[33]

Allan's discovery of the paper by the young Austrian mathematician, who was not yet 30 when this seminal work was published during the latter stages of World War I,[34] rekindled his own interest in the line integral problem and its application to radiological physics. A second reason why the summer of 1971 was a critical period for Allan was the receipt of a letter from Michael Goitein, requesting permission to use Allan's 1964 data in a new algorithm which he, Michael, had developed to reconstruct three-dimensional densities from two-dimensional projections.

Goitein had been a postgraduate student at the Harvard Cyclotron Laboratory, pursuing a PhD in particle physics, and met Allan from time to time in the lab although their research interests did not overlap. In 1969 Goitein accepted a post-doctoral fellowship at the Lawrence Radiation Laboratory in Berkeley and joined the biophysics group headed by Cornelius Tobias. Three years earlier, when Allan had been on sabbatical at Berkeley, he received a note from MR Raju, who worked under Tobias, requesting a reprint of Allan's paper on multiple scattering. Allan not only sent him a copy of this paper, but also enclosed reprints of the 1963 and 1964 papers on his line integral problem and, for good measure, sent reprints to Tobias as well.[35] Goitein had decided to change fields, from particle physics to medical physics, and approached Tobias for advice. While Tobias did not hold out any hope for employment, he provided Goitein with a few challenging problems, one of which was how to compute the densities within an object by taking pictures of it from many different directions.[36]

Goitein made a quick mental calculation, told Tobias he knew how to solve the problem and that he would be back in a few days to demonstrate the solution. He proposed a matrix inversion technique, originally thinking that an object of 80 x 80 pixels could be reconstructed by inverting an 80 x 80 matrix. After further consideration, however, he soon realised that he actually needed to invert a 6400 x 6400 matrix, a task well beyond the capabilities of computers in the early 1970s. Undeterred, Goitein came up with an iterative least-squares solution.[37]

In order to test his algorithm he needed some data, and this is when he wrote to Allan. Allan replied on 15 June 1971, sending Goitein a large box of IBM punched cards with a cover note explaining how the data had been gathered, with the final comment, "I hope that you can use this stuff!"[38] Goitein's algorithm not only worked well with Allan's data,[39] but he was also able to reconstruct the density of a chest phantom from data gathered by John Lyman using alpha particles produced by the LRL's 184-inch cyclotron. While his method had the dubious distinction of being one of the slowest reconstruction algorithms, it was also extremely accurate and flexible, accommodating different beam geometries. He submitted his manuscript to *Nuclear Instruments and Methods* in February 1972 and it was published later that same year,[40] and included a reconstruction of Allan's phantom (Figure 6.4). Although he was not offered the job by Tobias, Goitein returned to Harvard where, over the following three decades, he distinguished himself in the field of radiotherapy physics.

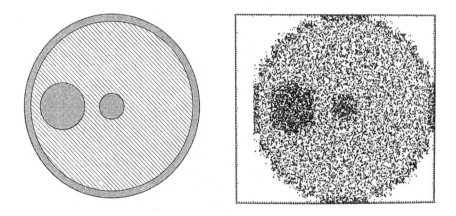

Figure 6.4   Michael Goitein's algorithm successfully reconstructs Allan's phantom, 1971.

A third discovery by Allan in the summer of 1971 was the work of his old friend and colleague, Aaron Klug. Klug was in England working at the Laboratory of Molecular Biology at Cambridge University. He and his colleagues had published a series of papers demonstrating how the three-dimensional structure of particles – such as the human wart virus, bacteriophages and the tobacco mosaic virus – could be reconstructed from their two-dimensional projections using an electron microscope.[41] They had developed a formal solution to the problem in terms of Fourier transforms and, using a least-squares approach, demonstrated how many separate views had to be included to yield a valid reconstruction to an acceptable level of spatial resolution.

On 15 December Allan wrote to Klug, describing his own work as a hospital physicist at Groote Schuur Hospital in 1956 and enclosed reprints of the 1963-64 papers on his line integral problem. He also included a draft copy of his manuscript that highlighted the work of Radon, requesting Klug's comments on it, and concluded with the statement:

> I am ashamed to say I was unaware that you had been elected
> to Fellowship of the Royal Society until I saw your article.
> How parochial can one become? Please accept my belated
> congratulations![42]

Klug responded immediately, acknowledging that he had been unaware of Allan's work or that of Radon, and enclosed a recent paper by his former student, Peter Gilbert.[43] Allan set about revising his own manuscript, in which he cited the work of both Michael Goitein and Aaron Klug, and concentrated on four separate applications: (1) determination of absorption coefficients for precise X-ray therapy; (2) scanning radiography using heavy charged particles, namely mono-energetic protons; (3) gamma-ray scanning; and (4) positron annihilation scanning (Figure 6.5).

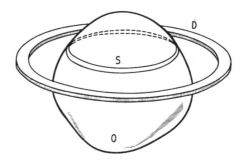

Figure 6.5   An object O contains positron annihilating material. An annular gamma-ray detector D defines a slice S of the object. The problem is to infer the density of the material from a number of different slices.

The paper was noteworthy for a number of reasons. First, it marked Allan's return to radiological physics, a field which he had all but abandoned since his 1964 publication in the *Journal of Applied Physics*. Second, this was the first time that anyone had cited the work of Radon in the medical literature. Third, Allan was able to cite, and therefore draw attention to, his own papers from 1963–64, publications that had garnered such a poor response from the scientific community. Finally, and most importantly, he chose to submit the paper to the journal *Physics in Medicine and Biology*, where it was most likely to be read by the appropriate medical audience. He sent the manuscript off in early February 1972; it was finally accepted after the third exchange of correspondence when the referee admitted she could not find Radon's article and Allan had had to send a copy to the editor.[44] The opening paragraph is a remarkably succinct statement of the scientific basis for computed tomography:

> The oldest problem in medical physics is as old as the discovery of X-rays, for an X-ray picture represents a projection of the matter in an object onto a plane. The interpretation of the picture is an attempt to reconstruct the spatial density from the projection. There are cases where ambiguities in interpretation require more detailed analysis. In such cases it is frequently necessary to attack the problem in a fully quantitative way.[45]

While Allan had been revisiting his line integral problem, and Michael Goitein had grappled with the challenge of inverting his large matrices, a British engineer was testing his own device for creating images of the human brain. On 20 April 1972, Godfrey Hounsfield and his clinical collaborator James Ambrose presented their findings to a meeting of the British Institute of Radiology in London.[46] *The Times* broke the story, featuring a photograph of a patient with his head located inside the system (Figure 6.6), while Hounsfield's employer, Electrical and Musical Industries (EMI) Limited, published a press release. In California, a fellow scientist burst into Goitein's office and said "Have you heard? Someone has announced a commercial version of the thing you've been working on!"[47]

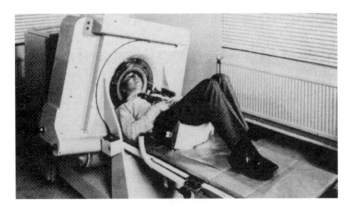

Figure 6.6   Hounsfield's original CAT scanner, released to the public in 1972.

The following month Hounsfield travelled to New York where, to great acclaim, he delivered a paper at a neurosciences meeting. The next day he appeared on all three major television networks in the United States.[48] Despite all the media attention, Allan Cormack was oblivious to the breakthrough by the British team. It was only after Goitein returned to Boston in late May that he shared the news with Allan, who immediately sat down and wrote a letter.

12 June 1972

Mr Godfrey Hounsfield
E.M.I. Central Research Laboratories
Hayes
Middlesex
England

Dear Mr Hounsfield,

In the E.M.I. News release of April 20th I saw an account of
your Computerized Transverse Axial Tomography system. I
have, for some time, had an interest in the computation side
of the problem which your system solves, as you will see from
the enclosed reprints.

First, let me say ''Congratulations.'' It is high time that
something like this was on the market.

Second, could you let me have more technical details than are
provided by news release, particularly of paragraph 2 of page
2? The reason I ask for this is that I think I have a way of
computing the densities from the data which is much more
economical of computer time than most of the methods
currently in use (Fourier transforms, the method of my reprints,
matrix inversions of one form or another). To assess whether my
method would work for your system I need to know what your raw
data are.

I hope that I am not asking you to reveal any technical secrets.

Thanking you in anticipation,

I remain,

Yours sincerely

A.M. Cormack

A.M. Cormack
Professor of Physics

There was no response to the letter.[49] It would be another seven years before Hounsfield made contact with Allan.

The development of the computerised axial tomography (CAT) scanner had been necessarily shrouded in obscurity, given the commercial potential of the system.[50] Hounsfield, who graduated with a diploma from the Faraday House Engineering College in London shortly after World War II, had joined EMI in 1951. Among other projects he worked on the design of an all-transistor computer and the application of pattern recognition techniques to radar for the detection of enemy planes and warships.[51] In the mid-1960s EMI had the good fortune to produce The Beatles' recordings and, flush with their commercial success, were willing to support Hounsfield's request to pursue research in pattern recognition applied to medical problems.

It was during this period that he came up with the idea of the CAT scanner. Because he worked for a commercial company rather than an academic institution, Hounsfield chose to protect his ideas by patent instead of publishing a journal article. He applied for his first patent in 1968 (UK number 1283915) and in it he referenced both of Allan's 1963–64 articles in the *Journal of Applied Physics*.[52] However, as Allan later acknowledged, "Hounsfield's algorithm (of the successive approximation type) bore no relation to any of mine".[53]

As an engineer, Hounsfield was not simply interested in conducting proof-of-concept tests. He wanted to build a device that physicians could use in the clinic to diagnose brain tumours. In addition to EMI's financial backing, he also secured further support – at first £2,500 and then later £12,000 – from the UK's Department of Health and Social Security.[54] At the Atkinson Morley's Hospital in Wimbledon, London, he met the neuroradiologist James Ambrose, with whom he established a very successful collaboration. Ambrose, who was South African, had been a Spitfire pilot during World War II and was renowned for his skill in diagnostic radiology. They tested Hounsfield's original CAT scanner prototype on preserved human brains from the local pathology museum, and then conducted a series of experiments on bovine brains. These animals were killed by cutting the great vessels of the neck, the kosher method performed by a *shohet*, which prevented haemorrhage in the brain.[55] From the images Ambrose could clearly identify vital brain structures: white and gray matter, the ventricles and the choroid plexus.

Their first clinical scan of a human was conducted, under a blanket of secrecy, on 1 October 1971. The patient was a 41-year-old woman with a suspected tumour in the left frontal lobe of her brain. Ambrose and Hounsfield were both astonished and delighted when the image clearly showed a dark circular cyst in exactly the location predicted by Ambrose's clinical tests. Within the next six months a further 70 head scans were performed; histologic evidence confirmed the existence of a lesion in almost every case. This was an impressive achievement given that each slice took six minutes to acquire and the data, recorded on magnetic tape, had to be transported across London for processing on Hounsfield's computer at the EMI laboratories.[56] The public presentation of their work in April 1972 launched the CAT scan era.

At the annual gathering of the Radiological Society of North America, held in Chicago in November 1972, EMI presented the images from Hounsfield's brain scanner, to great acclaim. Orders came pouring in, and the high cost – $300,000 – was not a deterrent. EMI did not realise the impact that their CAT scanner would have and estimated the worldwide demand to be no more than 25 machines.[57] In contrast, other manufacturers of medical imaging equipment soon recognised that there would be an enormous demand for the technology. By mid-1976 there were more than 20 companies making CAT scanners; the devices could now incorporate the whole body, and the cost had risen to half a million dollars. Worldwide sales exceeded 1,000 scanners. Bettyann Kevles, who wrote a history of medical imaging entitled *Naked to the Bone*, concluded:

> The combination of The Beatles' success with the British system of research subsidies, and the genius of one engineer, broke the cash barrier and changed the face of modern medicine.[58]

In 1970 two mathematicians working at the State University of New York in Buffalo published a new algorithm for reconstructing three-dimensional objects based on data from electron microscopy and X-ray images.[59] Richard Gordon and Gabor Herman, affiliated with the Center for Theoretical Biology and the Department of Computer Science, presented their algebraic reconstruction technique (ART) and claimed that it had six distinct advantages over the Fourier technique developed by Aaron Klug and his colleagues at Cambridge University.[60] Klug was unimpressed. He and Tony Crowther wrote a critique for the *Journal of Theoretical Biology*, pointing out the shortcomings of ART.[61] In a letter to Allan in January 1972, Klug was far more forthright:

> It shows that the criticisms by Crowther and myself were too mild, though we did guess the secret of their success with the model examples, namely using the same rules to generate the projections as were used in solving the equations.[62]

By 1973 the controversy had become quite heated and had attracted attention in the general scientific press. The National Science Foundation, which had provided considerable funds to the ART team, was anxious to achieve some form of resolution. Bob Marr, working in the Applied Mathematics Department at the Brookhaven National Laboratory on Long Island, New York, was prevailed upon to organise a workshop.[63] He created a steering committee and invited Ron Bracewell, Allan Cormack, Richard Gordon, Aaron Klug, David Kuhl and Paul Lauterbur to assist him in setting up the programme.

Attendance was by invitation only. When word of the meeting first leaked out, Marr was besieged with phone calls from around the country. He was told by a colleague from Berkeley that he had the "hottest thing going" that summer. Gordon, who had left Buffalo and was now at the National Institutes of Health in Bethesda, outside Washington, DC, had just published a comprehensive review of the field with Herman,[64] and so was familiar with the work of many authors. His advice was invaluable in choosing which speakers to invite. For four days, from 16 to 19 July 1974, 59 scientists gathered at Brookhaven for an international workshop entitled 'Techniques of Three-Dimensional Reconstruction'.[65] They not only came from all over the world, but they also represented diverse scientific fields such as nuclear medicine, electron microscopy, radio astronomy, holographic interferometry and plasma physics.

Conspicuous by his absence was Godfrey Hounsfield, although Bernard Mayo of EMI's Central Research Laboratories in England did attend and presented a theoretical paper on tomographic reconstruction using Walsh functions. For Allan, this proved to be a remarkable workshop for five important reasons. First, he renewed his friendship with Aaron Klug whom he had not seen for over 20 years. Second, he made the acquaintance of Ron Bracewell, who had published a solution to the line integral problem in the mid-1950s – at about the same time Allan encountered the problem at Groote Schuur Hospital – with application to radio astronomy.[66] Surprisingly, they had not known one another in the late 1940s when both were studying at Cambridge and Bracewell, as an Australian, was friendly with Joan Freeman, Allan's fellow Nuclear Nit-Wit.

Third, although Allan did not present a paper at the workshop, he chaired a scientific session on 'Theory, Algorithms and Data Analysis'. This pitted Klug and Crowther from the Laboratory of Molecular Biology in Cambridge against Herman and the proponents of ART. Although their differences were resolved amicably, Allan saw first hand just how much was at stake in the contest. Fourth, he was convinced that Radon's transform was very robust, meaning that almost any method of attack with it would yield useful results. Finally, for the first time Allan met Lauterbur, whose paper on 'zeugmatography' and nuclear magnetic resonance (NMR is a physical phenomenon based on the quantum mechanical magnetic properties of an atom's nucleus) left a lasting impression:

> His paper on NMR imaging left me with a queasy feeling about the future of CAT scanning with X-rays, since with NMR several parameters could be measured, all potentially interesting, whereas with X-rays there was only one: attenuation.[67]

Lauterbur was then working at the State University of New York in Stony Brook, also on Long Island and located just a few miles from the Brookhaven National Laboratory in Upton. He and Bob Marr toyed briefly with the idea of a follow-up meeting until they realised the first one had been so successful that another meeting was unnecessary: "Like the proverbial six blind men, we had all recognised the elephant together".[68] It would be almost three decades before Lauterbur was awarded a share of the Nobel Prize in Medicine for his seminal contributions to the development of magnetic resonance imaging (MRI).

In his 1963 paper in the *Journal of Applied Physics*, Allan had mentioned that protons, instead of X-rays, could be used with his line integral solution to measure the density of the human body. He did caution, however, that the problem was complicated by the fact that bone and soft tissues had variable chemical compositions and this would influence the rate of energy loss of the protons. Despite these misgivings, a seed had been planted. Five years later, his good friend Andy Koehler conducted a series of experiments at the Harvard Cyclotron Laboratory, creating the first proton radiographs ever produced. His image of an aluminium absorber was published in a ground-breaking paper in *Science*,[69] while his first radiographs of biological tissue, featuring a lamb chop, astonished his colleagues at the laboratory.[70] Although the spatial resolution was inferior to X-rays – the fuzziness caused by the scattering of protons – the technique had two major advantages over X-rays. First, the ionising radiation, or dose, to which the patient would be exposed was considerably lower. Second, the Bragg effect – the sharp fall-off in protons at a particular distance as they penetrated the tissue – meant that very small variations in density could be detected.

Koehler's work caught the attention of Bill Steward, a pathologist working at the University of Chicago. He immediately recognised the potential of proton radiography in the detection of tumours, cancerous lesions that often have a density very similar to that of the surrounding tissue. While such lesions were typically invisible on an X-ray image, Steward was hopeful that they might show up on a proton radiograph. He brought to Boston autopsy specimens of human brains, fixed in formalin, with a variety of lesions, including tumours and cerebral infarctions. The results of the experiments they conducted on his specimens demonstrating the superiority of protons, were published in *Nature*.[71] In their next set of experiments, Steward and Koehler went one step further. A live patient with carcinoma of the breast was imaged with protons and the diagnosis confirmed.[72]

Although he was not an active participant in their collaboration, Allan was aware of the research being conducted on proton radiography. In November 1973 he wrote to Steward, describing his own background and experience with the cyclotrons at Harvard and Berkeley. He mentioned his research at Groote Schuur Hospital in 1956 and enclosed reprints of his 1963–64 papers, describing them as 'pioneering', and also included his recently published paper on the Radon transform. Allan commented on the recent introduction of the CAT scanner:

> The reconstruction problem of X-rays has become fashionable in the last few years, a commercial scanning machine is now on the market, and there seems to be little more to be done at a fundamental level even though a number of technical problems remain unsolved. Heavy charged particles, for example protons, are quite another matter.[73]

Allan Cormack had conceived a bold idea: proton tomography. He believed it should be possible to employ proton beams for two purposes: first to determine the location of a tumour, and then to treat it. The pursuit of this idea was to consume much of his time and creative energy over the ensuing six years.

In January 1975 Allan was approached by Rodney Brooks, working as a biophysicist at the National Institutes of Health (NIH) in Bethesda, Maryland. Brooks had taught in the summer school at Tufts in the early 1960s while pursuing a PhD in physics at Harvard, and was now working for Giovanni Di Chiro, head of neuroradiology at the National Institute of Neurological Diseases and Stroke.[74] Di Chiro was co-author of a paper that had recently appeared in *Science* describing a whole-body tomographic instrument that was called the Automatic Computerized Transverse Axial (ACTA) scanner. The lead author was Robert Ledley of Georgetown University in Washington, DC, and they drew specific attention to Allan's work:

> Cormack, in a remarkable paper of 1963, discussed the idea of making measurements of the X-ray transmission "along lines parallel to a large number of different directions" so as to obtain a sequence of X-ray transmission profiles.[75]

Shortly after Brooks arrived at the NIH, Di Chiro asked him to write a review article on computed tomography and so he became familiar with Allan's 1963–64 papers as well as the more recent work on the Radon transform. Brooks sent a draft copy of the article and requested permission to include a photograph of Allan's prototype CAT scanner which, remarkably, had never before been published. He also indicated that Di Chiro was interested in exploring further developments that might improve the art of image reconstruction.[76] During a series of follow-up telephone conversations, Allan described the work that he and Andy Koehler had been planning on proton tomography.

With the encouragement of Di Chiro and Brooks, Allan and Andy conducted a set of experiments at the Harvard Cyclotron Laboratory in the summer of 1975. Using very simple equipment, they made measurements on a circularly symmetric phantom consisting of Lucite and a range of sugar solutions. Their findings, later published in *Physics in Medicine and Biology*, demonstrated that density differences of just 0.5 percent could be detected and reconstructed with reasonable accuracy.[77] Brooks had the NIH engineering workshops build him a special electronic detector and then spent a few days in Boston doing experiments at the cyclotron. While he was there, he encountered Allan's acerbic sense of humour. The proton therapy treatments were being conducted and one of the patients developed a medical problem. Knowing that the building was filled with PhDs, Allan called out "Is there a *real* doctor in the house?"[78]

The summer was also a time when the Cormack family returned to their vacation retreat in Alton, New Hampshire. By now Robert was 10 years old and Allan built him a tree house in a pine tree near the road that led to the family homestead. He also introduced his son to the mysterious possibilities of physics. On a clear sunny day, Allan accompanied Robert to Lake Winnepesaukee and showed him how to use a small mirror to transmit a signal by means of reflected light.[79] Then, leaving his son behind with the mirror, Allan climbed Mount Major with his own mirror. Between the pinnacle of the mountain and the shores of the lake, father and son established a simple form of communication.

Allan also took Robert to his physics laboratory at Robinson Hall, where his job was to assist with reading the counters for an experiment his father was conducting. This required him to climb to the back of the lab to take the readings while his father fed the data into the physics department's minicomputer, a PDP-11, manufactured by the Digital Equipment Company (DEC) of Boston.[80] These early opportunities for Allan to share his passion for physics with his son undoubtedly had an influence on Robert's own selection of an undergraduate major when he entered Harvard eight years later.

It had been over eight years since Allan returned from his year's study leave at Berkeley, and he would soon be stepping down as chair of the physics department at Tufts. It was time to consider another sabbatical. On 20 October 1975 he wrote to his old colleague Godfrey Stafford, who by then was director of the Rutherford High Energy Laboratory in Berkshire, England. Allan enclosed a preprint of his article with Koehler and outlined the requirements for enabling proton tomography to detect density differences at not just the tenth percent level but at the hundredth, or possibly even thousandth, percent level. His enthusiasm for the project came bubbling through:

> There is a great deal at stake here. If protons can do a better job than X-rays, the accelerators and detectors will sell like hot cakes, the way the EMI-scanner is selling. Financial matters apart, there is a question of prestige. Physics is under pressure to show that it is 'useful'. I believe, and I suspect that you do

too, in physics for its own sake. When I use the word 'prestige', I mean the prestige that will accrue to someone (who the public sees as a purist) not only solving 'useful' problems, but able to solve them with his strong hand tied behind his back.[81]

Three days later Allan was visited by a group of engineers from the Gordon Engineering Company, who were based in Wakefield, Massachusetts.[82] The group's leader was Bernard Gordon, who had spent a period at Tufts in 1944 as part of the US Navy's college training programme. As a creative engineer and entrepreneur, he had invented and commercialsed the first analogue-to-digital converter, and his company was about to build its first CAT scanner. They were looking to improve its performance and sought Allan's advice.[83]

Allan stood at the blackboard in his office, sketched a diagram of each component of the machine and then explained its function. By reducing the profound mathematics to the level of high school geometry, he soon made the functions crystal clear to the attentive engineers. From this interaction, Gordon and his colleagues were able to invent the instant imaging CAT scanner, a state-of-the-art technology which they licensed to Siemens.[84] Over 30 years later, the company – now named the Analogic Corporation – is still in the business of manufacturing scanners and other medical imaging systems.

Shortly after Bernard Gordon's visit, Allan heard from Giovanni Di Chiro. The neuroradiologist emphasised that he was committed to the project on proton tomography as long as his institute in Bethesda had something to show for its involvement with the project. While acknowledging that it would be a preposterous pretence to suggest he could provide technical support in the early stages, Di Chiro was confident that, later on, his guidance would be of substantial help in producing a functional machine. He pointed out that the success of the EMI scanner depended not only on the technical expertise of Godfrey Hounsfield but also on the clinical insights and advice of James Ambrose. His closing comment that "Incidentally, Hounsfield is currently being considered for the Nobel prize"[85] must have given Allan pause for thought, to wonder if he too was perhaps a candidate for this accolade.

Chapter 7

On the Road to Stockholm

There is irony in this award, since neither Hounsfield nor I is a physician. In fact it is not much of an exaggeration to say that what Hounsfield and I know about medicine and physiology could be written on a small prescription form!

Allan Cormack[1]

In mid-November 1975 Allan came across a recent article by James Ambrose, entitled 'A brief review of the EMI scanner', which had just been published in the *British Journal of Radiology*. Ambrose provided a short introduction to Hounsfield's research on pattern recognition and then speculated that an assiduous search of the literature must have been made at the time. Ambrose thought that Hounsfield may or may not have been aware of the work of Oldendorf (1961), Cormack (1963), Cameron and Sorenson (1963), and Kuhl (1968).[2]

After reviewing the four papers by these authors, Ambrose went on to describe his own collaboration with Hounsfield, first their work on brain specimens preserved in formalin, then fresh animal brains, and finally the study of patients with suspected brain tumours. Allan sat down and wrote a letter to the editor in the form of a brief technical note entitled 'More on the EMI scanner', in which he acknowledged that Ambrose had correctly described the simple experiment he had performed at the University of Cape Town in 1957. He also outlined, in some detail, the key features of his 1964 paper which Ambrose had not cited and finished off the letter with the statement:

> I suggest that, apart from a different reconstruction algorithm
> (of which there are now many), the EMI scanner is a finely
> engineered implementation of the principles and practice set
> forth in my papers.[3]

The managing editor, Mrs L Surry, acknowledged receipt of the letter
and provided him with Ambrose's address.[4] Allan immediately wrote to
Ambrose, providing him with a copy of the technical note, a reprint of his
1964 paper and a photograph of his prototype scanner.[5] In early January
1976 Ambrose responded, thanking Allan for the enclosed materials, admitted that he was unaware of the 1964 paper, and gave the assurance that
his intention was not to be controversial or knowingly to have omitted any
contributory work.[6] A week later Dr David Bewley, honorary editor of the
British Journal of Radiology, followed up with Allan, saying that he did
not wish to start a troublesome correspondence in the journal. However,
Bewley did conclude by offering to publish Allan's technical note provided
he did not claim that his papers provided the sole basis upon which the
EMI scanner had been developed.[7] Sensing that his note, if published in the
journal, might be perceived as self-promotion, Allan let the matter drop.

Allan clearly realised that, in addition to his 1963–64 papers in the *Journal of Applied Physics*, he needed physical evidence to substantiate his work
at Groote Schuur Hospital and the University of Cape Town in 1956–57.
In early December 1975 he sat down and wrote to Robin Cherry and John
Juritz.[8] He enclosed reprints of the two papers and asked if his old UCT
colleagues could try and locate the two lead collimators plus the circularly
symmetric phantom made from aluminium and oak. Given their historic
significance, and sentimental value for Allan, he asked if photographs could
be made. Cherry responded a few weeks later, saying he was aware of
Allan's pioneering work because he had read the complimentary remarks
made by Ledley in *Science*.[9] He also informed Allan that the physics department had moved to the new RW James building in 1966 and speculated
that the apparatus must have been thrown out at that time. Cherry and
George Laing had searched through the laboratories and store rooms but
to no avail.

At about this time Allan mentioned to Kathryn McCarthy, his colleague
in the physics department at Tufts, that Hounsfield was being considered
for the Nobel Prize. McCarthy, who by now had been promoted to Provost
at Tufts, passed this information on to Burton Hallowell, the President.
Hallowell then asked Allan to summarise his involvement with tomography, and on 10 February 1976 he completed a five-page document entitled

'Chronology of my Connection with Computer Assisted Tomography'.[10] His chronology began with his stint at Groote Schuur Hospital in 1956 under the mentorship of Dr J Muir Grieve, and finished with his most recent collaboration on proton tomography with Andy Koehler, Rodney Brooks and Giovanni Di Chiro. Two weeks later Hallowell wrote to Salvatore Luria, a biologist at the Massachusetts Institute of Technology and 1969 winner of the Nobel Prize in Medicine for his work on the genetic structure of viruses. Hallowell enclosed copies of Allan's chronology, his curriculum vitae and reprints of his 1963–64 papers. He mentioned that Hounsfield, the designer of the EMI scanner, was being considered as a Nobel candidate and made the case for Allan:

> His contribution to tomography has caused a significant revolution in diagnostic medicine. We are writing to ask your evaluation of Allan's work and your advice as to directions along which we might proceed. Either I or the Provost Kay McCarthy will call you in two or three weeks.[11]

Nomination for the Nobel Prize in Physiology or Medicine can only be made by certain people. These include members of the Nobel Assembly at the Karolinska Institute in Stockholm, members of the Royal Swedish Academy of Sciences, medical professors at Scandinavian universities, other professors around the world selected by the Assembly, and Nobel Laureates in Physiology or Medicine. Luria was obviously qualified to nominate Allan but it is not known if he did. The statutes of the Nobel Foundation restrict disclosure of information about the nominations, including the names of the nominators, for 50 years. What is known is that Allan could not have been considered for the award in 1976 since the deadline for submission of applications in any year is 31 January.

In early April, Allan flew to San Juan, Puerto Rico, where he attended an international symposium on computed tomography. He presented a paper on proton tomography, co-authored by Koehler, Brooks and Di Chiro, in which he provided a progress report on their efforts to detect density differences of 0.1 percent or better.[12] Their paper generated considerable response, particularly from two professors of radiology affiliated with the Karolinska Hospital in Stockholm. Torgny Greitz and Björn Nordenström approached Allan after his lecture to discuss the clinical implications of imaging with protons. They also expressed an interest in his early work on computer assisted tomography.

On 28 April, Allan wrote to Greitz and included a preprint of his paper with Koehler on proton tomography, plus reprints of his 1963–64 and 1973 articles.[13] He also enclosed a copy of the chronology document he had prepared for Hallowell in February. Allan promised to send good quality proton radiographs of the breast as soon as he received them from Bill Steward in Chicago. These prints were duly posted off to Greitz and Nordenström in early June, although Allan was not entirely satisfied with the quality of the images.[14]

The review article on the principles of computer assisted tomography by Brooks and Di Chiro appeared in *Physics in Medicine and Biology* in May. At 44 pages, it provided a comprehensive appraisal of the field, including its historical development. The authors pointed out that the introduction of the EMI scanner was preceded by the work of Takahashi, Oldendorf, Kuhl and Cormack:

> These early attempts at image reconstruction used variations of Back-Projection and hence produced highly blurred images. A significant step was made by Cormack (1963, 1964) who developed a mathematical technique for accurately reconstructing images from X-ray projections and applied it to measurements on simple phantoms (fig. 4). Cormack's first studies were performed at the University of Cape Town, South Africa, in 1957.[15]

Figure 4 was the photograph of Allan's prototype CAT scanner (see Figure B.3 on page 227) that had never been published before. Although the caption incorrectly described the phantom as an aluminium cylinder surrounded by a wooden annulus (this was the phantom he had used in his Cape Town experiments in 1957), the impact of this article was crucial. Anyone researching the history of computer assisted tomography – such as the five-member Nobel Committee that screens the nominations and selects the candidates for the Physiology or Medicine prize – could not have missed the fact that Allan had conceived of a system with two key attributes. First, he had built an apparatus that incorporated both linear and angular scanning motions; and second, he had developed a mathematical algorithm for accurately reconstructing images.

Though a sabbatical at Godfrey Stafford's laboratory did not materialise, Allan and Barbara nevertheless decided to take their family on a month-long vacation to the United Kingdom. They extended the mortgage on their home to finance the trip, and on 15 June 1976 the family boarded a British Overseas Airways Company (BOAC) flight at Boston's Logan airport. Margaret, who by now had graduated from Harvard with a degree in literature and Germanic languages, was then enrolled for a diploma in medieval studies at the University of St Andrews in Scotland. She planned to meet them when their plane landed in Glasgow. During the flight over, Jean, in her second year majoring in music at Tufts, suffered from motion sickness. She asked for a glass of Coca Cola, which she knew from previous experience had helped to treat the condition.[16] The air hostess insisted that she have a cup of tea instead and Allan was hard-pressed to convince his younger daughter to follow this strange British tradition!

Shortly after their arrival on the west coast of Scotland, the family rented a car and set out for Caithness in the north. Their first stop over was at Crieff, just west of Perth, where they visited the town's museum. Young Robbie, not yet 12, was fascinated to discover a bronze stencil for 'Cormack's Herring'. He learned that this seal, used to adorn the barrels of herring, had been issued by the Scottish government to his great-grandfather, George Cormack, for his pickling process that was different to the traditional method.[17] The tour guide explained to the family that the Russians had preferred the Cormack herring and apparently asked for it by name. Years later, when Margaret was visiting friends in Iceland, she recalled this story. They laughed and told her, "Everyone knows the Russians love rotten fish!"[18]

Allan took great pleasure in showing his family the places he had visited as a teenager in 1938 – Loch Ness, Wick, John O'Groats, the Orkneys, Ullapool and Ballachulish – and introducing his children to their Scottish heritage. Unfortunately his mother's younger siblings, Aunt Betty and Uncle John, had both died in the early 1970s and Allan believed that any connections to relatives living in Scotland had been irrevocably lost.[19]

While the Cormack family was travelling around the United Kingdom, the United States Bicentennial was celebrated on Sunday 4 July 1976. Commemorating the adoption of the Declaration of Independence, the celebrations gave rise to feelings of nostalgia as a wave of patriotism, adorned in the national colours of red, white and blue, swept across the nation. It was, especially, a time of reflection. President Gerald Ford had commissioned the National Science Foundation (NSF) to comment on the state of scientific research, and Allan, as outgoing chair of the physics department at Tufts, had been asked to contribute. He identified just one problem which, in his mind, transcended all others:

> This is the erosion of the traditional view of what the function of a university is. I see this in the population at large, in many of their elected representatives, in many federal bureaucrats (even in the NSF), and, alas, in many university administrators and

students. A university should be a place where scholars congregate to pursue freely the intellectual problems which interest them. In return for this freedom, the scholars pass on their knowledge and stimulate the intellects of their students. My colleagues and I feel that the people named above have, each in their own way, demanded that we explicitly demonstrate in our work innovations, relevance and concern for interdisciplinary matters to an extent that we have lost much of what is most valuable in solving any problem – the time to think.[20]

As an aspiring scholar in her own right, Margaret had some understanding of the burden her father had carried as chair of the physics department. She also learned the lesson that if you did not participate in administration, you had no right to complain. During Allan's last few years as head of department she saw him becoming grumpier and grumpier. Margaret thought it was just old age, "But as soon as he stepped down, his relaxed attitude and acerbic sense of humour resurfaced".[21]

Within a few weeks of the Cormacks' return from their summer holiday in Britain, Allan completed a short four-page article with Brian Doyle, one of his postgraduate students. Their algorithm was an application of the 'hole' theorem to the Radon transform, where the emphasis was on filling in when the data were incomplete, a process which they referred to as 'peeling the onion'.[22] Having submitted the manuscript to *Physics in Medicine and Biology*, Allan also sent off copies to Aaron Klug at Cambridge University and Ron Bracewell at Stanford since he had cited their publications. He asked them if they could find any errors in the arguments presented. Klug, who by now had had a brief collaboration with Godfrey Hounsfield as both were Fellows of the Royal Society,[23] said he thought the result was mathematically correct.[24] Bracewell was even more enthusiastic:

> I enjoyed reading your note on algorithms for two-dimensional reconstructions. It seems to me that you are absolutely right. The idea is most intriguing and is bound to generate a lot of activity. It's a stimulating paper.[25]

Unfortunately for Allan, his two esteemed colleagues had missed critical errors in his mathematical reasoning. While Bracewell was correct in surmising that the paper would lead to some debate, Allan could not have guessed where that would lead or the embarrassment the Doyle article would later cause.

For five days, from 11 to 15 October 1976, Allan attended an international symposium on computer assisted tomography at the National Institutes of Health in Bethesda, Maryland. Giovanni Di Chiro served as chairman and organiser and chose the theme of non-tumoural diseases of the brain and spinal cord because these presented significant challenges for investigators. He also arranged for a panel of pioneers to address the audience, which was treated to rare and humorous insights into the development of CAT scan technology.[26] The pioneers included William Oldendorf, Allan Cormack and Ron Bracewell (Figure 7.1). A fourth pioneer, David Kuhl of Philadelphia, who had made significant contributions to emission tomography with isotopes in the early 1960s, was scheduled to be one of the panelists[27] but was unable to travel to Bethesda. He had also been on the organising committee for the meeting at Brookhaven two years earlier but been unable to attend and, as a result, he and Allan were destined never to meet.[28]

Figure 7.1   Panel of pioneers (left to right): William Oldendorf, Allan Cormack and Ron Bracewell listen to Giovanni Di Chiro reminisce about his experience with computer tomography, October 1976.

The pioneers' panel was the first time Allan met William Oldendorf. He was familiar with the neurologist's highly cited publication of 1961, describing a method to detect discontinuities in density,[29] and was interested to learn why it was Oldendorf had shared the prestigious Lasker Award with Hounsfield in 1975. These awards had been made annually since 1946 for major contributions to medical science and were often referred to as 'America's Nobels'. Although Allan recognised the contributions of his fellow panelist, he also came to realise that Oldendorf did not understand the mathematics of CAT scanners or the critically important role of an algorithm for generating a clinically useful image.[30]

Meanwhile, remaining true to his conviction that scholars should be free to pursue problems they found to be intellectually stimulating, Allan continued to explore the applications of high energy protons to imaging. Beginning in late 1976, and extending over the next two years, were his 'Archaeological Adventures with Andy'.

In March 1977 Allan and Andy Koehler attended an international symposium on archaeometry held at the University of Pennsylvania's museum in Philadelphia, where they presented a paper on the use of protons for nondestructive testing of materials.[31] They demonstrated how protons were superior to X-rays when measuring chemical composition in thick objects, illustrating their talk with a proton radiograph of a coin that clearly showed the outline of Queen Elizabeth's head and shoulders.[32] This exposure to professional archaeologists presented them with several opportunities to apply their techniques to problems of artistic and historical interest.

John Slocum, a private coin collector from Newport, Rhode Island, left them with a valuable set of 16 coins from the County of Edessa in the Middle East that dated from the time of the Crusades (12th century AD).[33] The problem was to try and identify the sequence of overstrikes on the coins. This information was important to establish the order of the rapidly changing rulers of Edessa in those turbulent times. As part of their experiments, Allan and Andy created very detailed moulds of both sides of the coins using silicone rubber, transferring the surface features onto a uniform material with a very different atomic number. Although unable to answer Slocum's question about the order of the overstrikes, they were able to measure the heights of the various letters on the coins.[34]

Another collaboration was pursued with Arthur Beal and Leon Stodulski of the Fogg Museum at Harvard. They started with a badly corroded piece of ancient bronze and, despite the corrosion, were clearly able to show a design incised into the bronze. One of their highlights was being able to demonstrate that a peculiar feature of a decorated Islamic bronze disk was not a rivet, as had been supposed, but was part of the original casting. This suggested that the bronze was a copy rather than an original.

Father Carney Gavin, curator of the Semitic Museum at Harvard, provided the two physicists with some of their most interesting challenges. The first of these challenges was a pair of Assyrian clay tablets that dated from about 1500 BC. These consisted of an inner clay tablet on which a document – typically a legal contract – was inscribed, totally encased in an outer clay envelope, also with inscriptions. Their task was to see if they could read the inscriptions on the inner tablet without breaking off the outer envelope. This was a tall order and, although they did not succeed, they learnt something new about stereoscopic proton radiography.[35]

Gavin also provided them with their most bizarre object, a small bronze Egyptian case that dated from 1200 BC. It was suspected that the case might contain a mummified snake but the question was how this could be verified without breaking open the case. A previous examination using a commercial CAT scanner based on conventional X-rays had produced a totally black image. Using proton radiography, Allan and Andy were able to reveal that there was definitely an object in the hollow interior of the case, although they could not say for certain if it was a snake.

With the encouragement of Dick Wilson from Harvard,[36] on 22 March 1978 Allan submitted a grant application to the Rowland Foundation, which was based in Cambridge, Massachusetts.[37] The two-year budget to support his research on proton tomography with Andy Koehler was set at a modest $40,294. Included in the application package was their image of the inside of the bronze Egyptian mummy case. Almost five months later, on 12 August, Allan received a brief letter from Philip DuBois, director of the foundation, saying that although the proposal had sufficient merit to be considered most seriously, they were unable to make a contribution to its support.[38] This was a bitter disappointment for Allan, particularly since he had such high hopes for proton imaging.

Allan subsequently discovered that his colleague in Chicago, Bill Steward, was co-investigator on a similar grant application submitted to the National Institutes of Health in November 1977. Ronald Martin, a scientist at the Argonne National Laboratory, was the principal investigator of the proposal entitled 'Development of a prototype proton CAT scan system' that had a two-year budget of $346,389.[39] A third scientist, Ken Hanson, was also a co-investigator. Their application was successful and the project commenced on 1 July 1978. Using the 205 MeV proton beam produced by their Booster I synchrotron, the Chicago team demonstrated significant dose reduction and improved density resolution compared to conventional X-ray techniques.[40] They also confirmed the significant differences in the proton stopping power of biological tissues and, therefore, the considerable potential for soft tissue imaging.

A major obstacle to proton tomography was cost: a cyclotron was considerably more expensive to build than an X-ray tube. Allan, however, did not see this problem as insurmountable. He envisaged that a large metropolitan area would have a 250 MeV proton accelerator with a number of separate ports, perhaps as many as ten, that would enable several patients with cancer to be treated simultaneously.[41] If one of these ports was dedicated to proton tomography, there would be a small marginal cost for this extra feature. The key to the success of Allan's proposal was the necessity for diagnostic radiologists and therapeutic radiologists to work together.

Why, then, did proton tomography never take off? There are three main considerations when evaluating any medical imaging technique: cost; image quality; and dose. In the late 1970s, computed tomography (CT) based on X-rays still had the edge over proton tomography in terms of cost and image quality. Rodney Brooks, Allan's collaborator from the National Institutes of Health, provided a succinct answer to the question:

> As CT developed, however, its image quality became very good
> with doses that were not deemed to be of concern, so that took
> a lot of wind out of the proton sails. If there was any wind left,
> it would have disappeared completely when MRI came along
> and produced remarkable image quality at zero dose.[42]

Godfrey Hounsfield was awarded his first United States patent on 11 December 1973,[43] and before the decade was out he had added a further ten patents.[44] In each case the patents were assigned to his employer, EMI Limited, based in England. EMI's first sale of a CAT scanner in the USA was in 1973 and by mid-1975 the company had installed 160 machines which, at an average cost of $400,000, represented a considerable volume of business. The success of EMI attracted other companies to the market, with sales forecasts estimated to be as many as 400 new machines per annum.[45] Not surprisingly, EMI sought to defend its dominant market position. Fearing a Declaratory Judgement (in which the court in a civil case declares the rights, duties, or obligations of each party in a dispute), they filed a patent infringement suit against their major competitor, Ohio Nuclear Incorporated, in Cleveland during 1976. Further suits followed over the next three years against the Picker Corporation, General Electric and Pfizer Medical Systems.[46]

EMI's legal representative in the USA was Ivan Kuvrakov, who became known as 'Ivan the Terrible' because he relentlessly pursued any application of X-rays that appeared to infringe on Hounsfield's many patents.[47] Allan was retained to serve as an expert witness for a few of the patent infringement suits and this gave him an interesting perspective on the process. According to United States law, there are sweeping 'discovery' provisions under which each party is obliged to disclose to the other parties all documents which might possibly be relevant. During the discovery process, reference to 'Allan Cormack' was disclosed in Hounsfield's notebooks for 1967 (the year before the first UK patent applications by EMI), as well as a London telephone number – 278 2890 – which turned out to be that of Denis Rutovitz.[48] By the late 1970s, Allan's old friend had moved from London to Edinburgh, and was contacted by American lawyers who were assiduously seeking to uncover anything that might have helped their clients' defence against EMI. Rutovitz was puzzled by their questions, since he had no recollection of ever having spoken to Hounsfield, but his subsequent letter to Hounsfield[49] left him no wiser as to the reason for his name and telephone number appearing in the notebook.

Allan believed that the genius of EMI's patent infringement strategy was to push its suits as far and for as long as it could without actually going to trial. By settling out of court, all the records would be permanently sealed, thereby preventing the public from gaining access to the arguments put forward by each side. A further advantage of settling the law suits *before* going to trial was that EMI avoided the possibility that prior discoveries by other scientists might have rendered Hounsfield's patents invalid. Their strategy paid off – Ohio Nuclear reputedly settled for $15 million[50] – and most manufacturers of CAT scanners throughout the world signed licensing agreements with EMI.

Having studied all the relevant patents in great detail, Allan acknowledged that Hounsfield's algorithm, of the successive approximation type, bore no relation to his own solution to the line integral problem.[51] As indicated in his letter to Hounsfield, he was also delighted that the Englishman had succeeded in developing a working CAT scan system for clinical use (see page 142). However, he took the EMI patent attorneys, AB Strong and RA Hurst, to task over their interpretation of the discovery process:

> On the one hand we have Oldendorf and others (who knows, possibly Röntgen himself!) who say in effect "Wouldn't it be nice to convert data from a series of X-ray pictures into a picture of a section through the body?" but who do not tell us how to do it. On the other hand, starting with Korenblyum *et al.*, and including Hounsfield and myself, there were those who had the same feeling but who provided algorithms for converting the data to the picture. Only the latter seem relevant to the discussion.[52]

Unfortunately the success of EMI's strategy in defending its patents was not matched by equal success in marketing and engineering, and by the 1980s the company was no longer in the CAT scan business.

By late March 1977 Allan had received the second round of reviews for his paper with Brian Doyle on the hole theorem. He was unhappy with the suggestion by one of the reviewers that he and Doyle provide a test pattern to demonstrate the impact of a 'hole' in the data on the amount of noise generated. Allan told the American editor of *Physics in Medicine and Biology*, Fearghus O'Foghludha of Duke University, that this was akin to the latest fad in teaching mathematics.[53] Children were persuaded of the truth of the theorem that the sum of the interior angles in a triangle equals two right angles by having them draw a lot of triangles, measure their angles, and then discover the sums to be scattered around 180 degrees. Allan was adamant that theorems should not be proved by looking at special cases. Unfortunately for him and Doyle, however, they did not have a proof for their application of Radon's transform to the hole theorem. In the end, O'Foghludha relented and the paper was published in September that year.[54]

No sooner had the paper appeared in print than two groups of researchers identified what they believed were serious problems with the logic. The first was Paul Moran of the University of Wisconsin in Madison, with his postgraduate student, Randy Brown.[55] The other was Robert Lewitt, a post-doctoral research fellow with the State University of New York at Buffalo. Lewitt had just joined Gabor Herman's medical imaging research group, having recently completed his PhD at the University of Canterbury in Christchurch, New Zealand.[56] His thesis supervisor was Richard Bates who, along with Terry Peters, had published a pioneering paper on the mathematics of computed tomography in 1971,[57] a publication that Steve Webb described as "prophetic" in his book on the history of tomography, *From the Watching of Shadows*.[58]

Lewitt not only wrote up a technical note which he later submitted to *Physics in Medicine and Biology*, but he also made direct contact with Allan. In the last four months of 1977 Lewitt sent six letters, some with detailed mathematical equations, in which he attempted to convince Allan that there was no valid proof for the hole theorem.[59] Allan countered with letters and further equations of his own, remarking that Lewitt's results did "not inspire me with confidence".[60] He also made contact with Brian Doyle, sending copies of the correspondence and commented that perhaps they ought to be taking Lewitt more seriously.[61] Allan encouraged his old student to consider writing up a proof of the hole theorem for publication.

In August 1978 Todd Quinto was appointed as an assistant professor of mathematics at Tufts, having just completed his PhD at the Massachusetts Institute of Technology.[62] Since his thesis had been concerned with a mathematical model of tomography and the Radon transform, he made contact with Allan soon after his arrival at Tufts. The established physicist and young mathematician quickly established a positive rapport which was to develop into a mutually beneficial collaboration over the ensuing 15 years. Allan asked Quinto to evaluate Doyle's proof of the hole theorem and in January 1979 he delivered his verdict: the proof was fatally flawed.[63]

Lewitt, in the meantime, was still trying to get his comments on the Doyle paper published in *Physics in Medicine and Biology*. His letters to O'Foghludha were ignored, so finally, on 31 August 1979, he wrote a letter to the British editor of the journal.[64] During the same period Allan had had similar problems with O'Foghludha regarding another manuscript, prompting him to ask the British editor, "Would you be good enough to light a fire under whoever needs to have a fire lit under them, and have them accept or reject my note forthwith?"[65] Notwithstanding the frustrations experienced by both Lewitt and Cormack, it would be some months before resolution could be reached.

On 29 April 1978 the Trustees of Tufts University voted to award the Hosea Ballou Medal to Allan MacLeod Cormack. Nominated by his colleague Kathryn McCarthy, who was then Provost of Tufts, Allan received the medal for "distinguished service to the University and the world community through scholarship and research".[66] Since the Ballou Medal was (and is) considered more prestigious than an honorary doctorate,[67] this was obviously a great accolade for Allan and his family. The Board of Trustees had established the Hosea Ballou Medal in 1939 in honour of its inaugural President and by the mid-1970s there had been ten distinguished recipients of the award. Among these were two renowned engineers: Vannevar Bush, an electrical engineer, inventor of the differential analyzer (an early analogue computer), and responsible for establishing the National Science Foundation; and Herbert Hoover, a mining engineer, graduate of Stanford University, and 31st President of the United States.

Allan had continued to be an interested observer of, but not yet a participant in, the medical work of Andy Koehler and his clinical colleagues in using the proton beam produced by the Harvard cyclotron to treat cancer patients. Research on the radiobiology of pi mesons – also known as pions – had been pioneered in the United Kingdom at the Rutherford Laboratory.[68] As a result, a team of scientists at the Swiss Institute for Nuclear Research in Villigen began to explore the use of pions for the treatment of patients with cancer. Sir Brian Pippard, Cavendish Professor at Cambridge University, raised questions about the social responsibility of this initiative since the cost of accelerators to generate these charged particle beams might limit treatment to the wealthy.[69] While Allan harboured some doubts as to whether pions would turn out to be an improvement on protons and alpha particles, he was absolutely convinced that accelerator medicine need not be expensive. He argued that even an old accelerator, such as the 30-year-old Harvard cyclotron, could be run inexpensively. He was familiar with one therapeutic procedure – reducing the function of the pituitary gland – that illustrated the significant benefits of proton therapy:

> Patients who are treated arrive at the cyclotron under their own steam and in their own everyday clothes, and they depart in the same way after about an hour, having suffered only minor discomfort. The alternative is hospitalisation and major brain surgery at a cost of more than two times the proton treatment. The people who had the intelligence and courage to investigate this treatment are to be congratulated for the humane and economical results they have obtained.[70]

While Allan had pursued his career at Tufts for over 20 years, Barbara had stopped working in 1964 when Robert was born. She then waited for him to start first grade in 1971 before resuming her career as a mathematical analyst, working half-time in the applied sciences department at Harvard. This was volunteer work, solving the equations for a system of coupled harmonic oscillators, and writing a FORTRAN program for comparison with a numerical solution of the equation of motion.[71] In mid-1979, by which time Robert was almost 15, Barbara secured a full-time position with Intermetrics Incorporated, a company that Allan described as part of the "military-industrial complex".[72]

At Intermetrics, located on Concord Avenue in Cambridge, Barbara focused on the problem of relative navigation for a community of military vehicles travelling on land, by sea and in the air. Each vehicle measured its position and velocity in three-dimensional space using its own navigational equipment. At the same time, the vehicles communicated with one another by means of radio signals. Barbara's task was to design a mathematical filter and apply it to the length of time taken for the radio signal to travel between members of the community of vehicles, in order to determine an optimal estimate of their relative positions on a Cartesian grid. Among the vehicles were an F-16 Fighting Falcon jet and a Trident submarine.

The solution to the equations was not coming out right and this frustrated Barbara. Then the penny dropped. She was working with a set of three-dimensional axes in which the z-axis was vertically up for the air force but vertically down for the navy![73] As soon as Barbara realised the difference in the conventions and made the necessary sign changes, the problem was solved.

Although Allan was extraordinarily committed to his career as a physicist, he also had a diverse range of interests and read widely. When he returned to the University of Cape Town from Cambridge University in 1950 he had had discussions with John Day, professor of zoology, about ecology.[74] Day's particular interest was the ecology of estuaries, which he studied while seated in a dinghy, having lost a leg in action during the war.[75] Inspired by Day, Allan became increasingly interested in the study of animals and what they might contribute to the understanding of man. His daughter Jean gave him a copy of *On Human Nature* by Edward O Wilson, shortly after it was published in 1978. In this book, which would win him the Pulitzer Prize the following year, Wilson argued that the human mind was shaped as much by genetic inheritance as it was by environmental factors.

An entomologist by training, with a specific interest in ants, Wilson was also the author of the landmark text *Sociobiology: The New Synthesis*, published in 1975, in which he argued that all animal social behaviour was governed by rules laid down by the laws of evolution. His theory proved to be seminal, but it was no less controversial, particularly in its application to humans. The theory challenged the doctrine of *tabula rasa*, which held that the human mind was a blank slate waiting to be informed by culture and the environment. The major opposition to Wilson's ideas came from within his own department at Harvard, from the biologists Stephen Jay Gould and Richard Lewontin. They accused Wilson of "sending human nature back to the concentration camps".[76] Allan had been aware of the 'molecular biology wars' at Harvard, as they played themselves out in public, but his allegiance lay on the side of Wilson, a man who believed:

> All people yearn to have a purpose larger than themselves. We are obliged by the deepest drives of the human spirit to make ourselves more than animated dust, and we must have a story to tell about where we come from, and why we are here.[77]

Thursday 11 October 1979 was like any other week day in the Cormack household. Barbara had left the house first, setting out for work in Cambridge, followed shortly by Robert, who walked to the Winchester High School. Since Margaret was away at Yale, preparing for her preliminary exams before embarking on the PhD thesis, only Jean and her father were still at the house. Allan's alarm had gone off and, since this was before the 'snooze' button had been invented, he went back to sleep.[78] Suddenly, the bedside telephone rang and Jean thought "Oh no, now Dad's really going to be mad!" The caller was a reporter from one of Boston's radio stations who gave Allan the astonishing news that he had been awarded a share of the Nobel Prize for Medicine. Life for the modest physicist would never be quite the same again.

Jean, who was in her senior year at Tufts majoring in music, and her father ate a hurried breakfast before leaving for the university. As they emerged from the house, a cameraman from the *Boston Globe* photographed Allan, and a journalist, Robert Cooke, requested an interview.[79] Allan, however, was scheduled to teach his introductory physics class that morning, and was running late, so he and his younger daughter hurried on towards the Tufts campus. When he entered the lecture theatre the 260 members of his class gave him a standing ovation and presented him with a big bundle of helium-filled balloons, accompanied by a congratulatory note.[80] A good friend of Jean's, a fellow violinist who played in a string quartet with her, was a member of that class. He had heard the news on the radio that morning and was so shocked he had fallen out of bed! At the request of the students, Allan lectured on his line integral problem (Figure 7.2).

The Office of Public Information at Tufts released a statement to the media and arranged a press conference for that afternoon. The venue was the Coolidge Room in Ballou Hall, the main administrative building located at the centre of campus. Hosting the event was Jean Mayer, President of Tufts, and also present were Barbara, Jean and Robert. Representatives of the local and national news media, as well as members of the academic and administrative staff at Tufts, crowded into the room. Allan brought his prototype CAT scanner across from Robinson Hall, placing it on a table where it occupied centre stage (Figure 7.3).

Mayer was understandably delighted with the recognition that both Allan and Tufts University had attracted:

> We are here to honour a man who is the best example I know
> of that pure research has enormous payoffs. The only research
> money he received was a $900 grant from the Atomic Energy
> Commission and some support from the university. This is a

Figure 7.2   Allan lectures to his freshman physics class on his line integral problem, October 1979.

measure of the quality of the faculty at Tufts, that a Nobel Prize winner is teaching a freshman physics course. I hope undergraduate colleges throughout the nation will take note of that.[81]

Allan, with his self-deprecating humour, commented that his work on the CAT scanner had really just been a hobby, and he highlighted the contribution of Radon in 1917. He also described to the audience how the idea had presented itself to him back in 1956 when he was working at Groote Schuur Hospital in Cape Town (Figure 7.4). When this story was reported in *The Tufts Observer* the following day, there was, ironically, another story about South Africa: Tufts University, under pressure from its students, had divested its stock in two public companies, including the banking firm JP Morgan, which refused to subscribe to the Sullivan Principles.[82] These principles, introduced by the Reverend Leon Sullivan, were designed to force American companies doing business in apartheid South Africa to provide equal opportunities for the advancement of their black employees in that country.

Official notification from Stockholm also arrived that Friday. A telegram signed by George Klein and Bengt Pernow, chairmen of the Assembly and Committee respectively, informed Allan that the Nobel Assembly of the Karolinska Institute had decided to award the Nobel Prize in Physiology or Medicine for 1979 jointly to him and to Godfrey Newbold Hounsfield for

Figure 7.3   Allan at a press conference, with Barbara and Jean seated on the couch,
October 1979

their development of computer assisted tomography. Stig Ramel, President
of the Nobel Foundation, sent Allan a letter informing him that his share
of the prize amounted to 400,000 Swedish krone (about $95,690). He also
invited Allan and his family to Stockholm to participate in the Nobel Week,
and to attend the prize ceremonies on Nobel Day, 10 December 1979. Ramel
asked Allan to send an answer as soon as possible by telegram.[83]

The news spread rapidly around the world. There were telephone calls
from Allan's friends Godfrey Stafford and John Wanklyn in Oxford, and his
family in Johannesburg and Cape Town. The headline in the *Argus* that
Friday 12 October read 'Nobel Prize ... from an idea born in Cape', and
the *Cape Times* announced 'Nobel for ex-Rondebosch boy'. *UCT News*, a
publication of the University of Cape Town, stated 'UCT graduate shares
Nobel Prize', while the newsletter of the Rondebosch Old Boys' Club led
with 'Those were the days'. Also getting in on the act were the two Senators from Massachusetts, Paul Tsongas and Ted Kennedy, who sent letters
of congratulation.[84] That same day Allan also received a letter from the
White House.[85]

Figure 7.4   Allan explains to the audience how his CAT scanner works, October 1979.

Professor Allan McCleod Cormack
Department of Physics
Tufts University
Medford, Massachusetts 02115

To Professor Allan McCleod Cormack

Congratulations! I know that all Americans join me in
applauding the tribute you have received from the Nobel
Assembly.

Untold numbers of persons stand to benefit – as have
thousands already – from the advanced diagnostic procedures
made possible by your work on the theoretical processes
underlying the development of the CAT scan. Your work has
meant not only a triumph for science but also a triumph for
humanitarian concerns.

It gives me great pleasure to salute you on this occasion and
to send you my best wishes.

Sincerely,

[signature: Jimmy Carter]

When official announcement of the prize was made at the Karolinska Institute in Stockholm on Thursday 11 October, it was an hour late, and the
press release given to reporters was written only in Swedish – not Swedish,
English, French and German as in previous years. This immediately led
to speculation that something had gone awry. Although the deliberations
of the five-person Committee and the 64-person Assembly are confidential
and sealed for 50 years, reporters in Stockholm pieced together a plausible
explanation of the riddle.

For some years the Nobel Prize in physiology or medicine had been
awarded to basic scientists for fundamental insights into the nature of the
genetic code and the mechanisms controlling its expression. Some in the Assembly saw these as advances in molecular biology rather than in medicine
per se. In 1979, the CAT scanner was apparently up against work done on
immunogenetics.

For the first time in more than a decade, a radiologist, Ulf Rudhe,
joined the Selection Committee. During the ensuing debate in the Assembly, where a strong backer of the CAT scanner was Allan's friend Torgny
Greitz, a compromise was reached. Instead of awarding the prize to three
candidates, as was allowed in the rules, the least 'basic' of the scientists
named by the Committee would be eliminated. That person was probably
the clinician William Oldendorf. "Anybody who goes into science expecting to win the Nobel Prize is about as realistic as a person going to Las
Vegas to get rich", said Oldendorf at the time. "But my feeling is that I
should be going to Stockholm".[86]

Although Oldendorf had secured a patent for his device in 1963, and
subsequently shared the Lasker Award with Hounsfield in 1975, his original paper published in *Transactions in Biomedical Electronics* in 1961 was
conspicuous for its lack of mathematics. There was no algorithm to explain

how his meter readings could be converted into an image.[87]

Another clinician who might have been considered for the award was David Kuhl, a radiologist at the University of Pennsylvania. He was particularly interested in the use of radioactivity and emission tomography as a diagnostic tool, and built a series of scanners. He later acknowledged that "a modification of our 1965 transmission scanner by using a more effective reconstruction algorithm" would have resulted in a better quality imaging system.[88] The exclusion of a third candidate was later confirmed in a letter that Allan received from Ulf Rudhe:

> The decision of the Nobel Assembly of 1979 to award the prize
> to you and Mr Hounsfield – and not to anybody else in the field
> – fills me with persistent satisfaction.[89]

Intriguingly, the credit for having first developed the CAT scanner may not belong to either Allan Cormack or Godfrey Hounsfield. In 1983, four years after the Nobel Assembly had made its historic decision, Harrison Barrett of the University of Arizona discovered the work of a group of Soviet scientists.[90] In the late 1950s, three researchers from Kiev – Tetel'Baum, Korenblyum and Tyutin – published a series of papers in Russian which demonstrated that they had solved the theoretical problem of computer tomography. They also implemented their results experimentally. Barrett tested their algorithm and found it worked well, if a little slowly. Mysteriously, no further evidence of their research in the 1960s could be found, which seems rather sinister in view of the conditions prevailing in the USSR at the time.[91]

John Clifton, the British editor of *Physics in Medicine and Biology*, eventually resolved the thorny matter of the Doyle paper, writing to Allan on 12 October and inviting him to put the matter right in the columns of the journal.[92] Despite having received the life-changing call from Stockholm the day before, Allan soon responded and drafted a letter for publication. Clifton then wrote to Lewitt, enclosing a copy of Allan's letter, and requested that he withdraw his note.[93] He also apologised for the "incredible muddle" in dealing with the note and assured Lewitt that the journal would do their best to ensure such a "shambles" did not again occur. Lewitt acceded to Clifton's request and Allan's letter was published early the following year:

> Section 3 of a paper by Cormack and Doyle (*Phys. Med. Biol.*
> 22 1977 p 994) is wrong except for the trivial case of circular
> symmetry. This section was based on a proof which has since

been found defective. One flaw in the proof was pointed out by
Dr RM Lewitt, and other fatal flaws were pointed out by Dr
ET Quinto. I am indebted to Drs Lewitt and Quinto for their
observations.[94]

Given his recently elevated status, it could not have been easy for Allan
to have acknowledged his mistake. However, as an academic scholar he was
committed to 'seeking the truth' and in this instance he admitted that he
had been unable to find the 'truth'.

Less than a week after Allan received his telephone call, two Harvard
physicists received similar calls. Sheldon Glashow and Steven Weinberg,
both aged 46 and products of the Bronx High School of Science and Cornell University, had independently contributed to a unified theory that
explained the mutual interactions of elementary particles. The Americans
shared their Nobel Prize in physics with the Pakistani, Abdus Salam, who
was affiliated with Imperial College in London and the International Centre for Theoretical Physics in Trieste, Italy. Salam, who had earned his
PhD in 1952 from Cambridge University, overlapped with Allan in the late
1940s when they were both members of St John's College. Harvard University arranged a celebration for Glashow and Weinberg to which Allan was
invited.[95] It was, by all accounts, a festive occasion during which Shelly
Glashow, when asked what the award meant to him, exclaimed "Now, at
last, I can afford to paint my house!"[96]

Appropriately, Giovanni Di Chiro and Rodney Brooks wrote an article
for *Science* in which they described the work of Cormack and Hounsfield.
They commented on the fact that it was just eight years since the first patient had been scanned at Atkinson Morley's Hospital in Wimbledon, and
that the CAT scanner had had "an unmatched impact on the radiological sciences".[97] Appearing at the end of November, the paper also drew
attention to Allan's discovery of the Radon transform and acknowledged
the contributions of William Oldendorf and David Kuhl. Di Chiro and
Brooks, however, emphasised the need for a mathematical solution to the
reconstruction problem.

At the invitation of Wilhelm Oldberg of the Royal Swedish Academy of
Sciences, Allan was asked to prepare a paper of his own, to be published in
Les Prix Nobel, the official publication of the Nobel Foundation. This would
also be the basis for his scientific presentation in Stockholm on Saturday
8 December. Three separate scientific journals – *Science*, *Medical Physics*,
and the *Journal of Computer Assisted Tomography* – sought and received
permission to reproduce the article. Allan included a discussion of the

Radon transform applied to the hole theorem and drew attention to the fact that he and Doyle had published a solution that turned out to be wrong.

Since Allan also wanted to highlight the potential of proton tomography, he made contact with Ken Hanson, who by now had moved from Chicago to the Los Alamos Scientific Laboratory in New Mexico. Hanson was understandably delighted that his work would be recognised in this way and sent Allan a series of slides.[98] These included images of a normal brain and a heart that had suffered a myocardial infarction, with X-ray computed tomography being compared to proton computed tomography (Appendix B).

Allan had also been informed that at the grand banquet, one winner for each Nobel Prize was required to deliver a brief address not exceeding three minutes. This protocol was followed, whether one, two or three people were awarded the prize. Hounsfield telephoned Allan to discuss the speech:

> Hounsfield asked me if I would do it as he was shy of such formal occasions. I agreed to do it. At some point in our conversation he said "I have just been reading your papers ..." and, since we were on the phone, I had no need to conceal a broad smile![99]

Interestingly, there was a precedent for Hounsfield's request. Prior to accepting the inaugural Nobel Prize for physics in 1901, Wilhelm Röntgen had panicked when told he was expected to give an acceptance speech and was unable to speak.[100]

Allan was not the only member of the Cormack family busy with preparations for the trip to Stockholm. Margaret had just passed her oral exams at Yale and returned from New Haven, Connecticut to her parents' home in Winchester, Massachusetts. She attended a course in modern Swedish being offered at Harvard and went out shopping with Barbara to purchase new outfits for her mother. When Margaret was at Radcliffe nine years earlier, she had heard advertisements on the radio for an upmarket though conservative store called Gertrude Singer's Fashion Studio located just off Harvard Square.[101] She never imagined that she would ever go shopping there!

When Barbara applied for leave from Intermetrics Incorporated, she was told that ordinarily six months' notice was required if she was planning to leave the country.[102] Fortunately, the company waived the rule, but they insisted that Barbara, or any member of her family, had to use an 'Embassy approved' doctor while in Sweden. Jean had just been diagnosed with a cyst on her spine but fortunately she was allowed to travel. On the evening of Monday 3 December 1979, the Cormack family – Allan, Barbara, Margaret, Jean and Robert – boarded a flight at Logan International Airport, destined to arrive in London the following morning.

Flight SK526 from London touched down at 3:10 pm on December 4 at Stockholm's Arlanda International Airport. When the Cormacks stepped off the plane, they were greeted by Torgny Greitz. Also there to meet them were Sture Theolin and his wife Marit, assigned to Allan and his family as their attendants for the duration of their stay by the Swedish Ministry of Foreign Affairs. After collecting their luggage they made their way outside where Lars, who was to be the Cormacks' personal chauffeur for the following ten days, was waiting with a long black Mercedes limousine.[103]

After a quiet evening together at their hotel, Allan and Barbara spent Wednesday 5th in Helsinki, while Margaret, Jean and Robert remained behind, exploring Stockholm with Lars. When their parents returned the following morning, Allan raved about the previous evening's dinner. Not only had they been given their own personal menus, hand-printed with wooden blocks, but the food was "like nothing I have ever eaten before". During this praise of the meal of a lifetime, Barbara sat very silently. When Margaret asked her if she had enjoyed the meal, her mother commented, "I couldn't forget what it was I was eating". The main course was reindeer![104]

On Thursday evening there was a gathering of all the Laureates and their families. It was the first time that Allan and Godfrey Hounsfield met face to face (Figure 7.5). The Englishman was a bachelor and so his sister (who reminded the Cormack family of Agatha Christie's Miss Marple) accompanied him to Stockholm. Allan shared with Hounsfield the short acceptance speech he had crafted for Monday night's banquet, and his fellow Laureate indicated that he was pleased with the content. The two younger Cormack children were surprised not to meet that year's winner of the Nobel Prize for Peace, Mother Teresa of Calcutta. Unfortunately it was not to be. The Nobel Committee for the peace prize was based in Norway and the ceremony was therefore due to take place in Oslo.

Allan attended a press conference at the Karolinska Institute on Friday morning, and that evening the Cormack family, together with Hounsfield, enjoyed a dinner at the Thielska Galleriat in Djurgården, hosted by the 5-member Nobel Committee for physiology or medicine. It was another memorable occasion, followed the next morning by their Nobel Lectures. These were delivered in the main auditorium at the Laboratory of Berzelius (an early 19th century Swedish chemist who invented the modern chem-

Figure 7.5   Allan Cormack and Godfrey Hounsfield enjoy a story told by Torgny Greitz, December 1979.

ical notation) in the Karolinska Institute. Allan spoke on 'Early two-dimensional reconstruction and recent topics stemming from it' (see Appendix B), while Hounsfield's presentation was simply entitled 'Computed medical imaging'.[105] Although Allan enjoyed the occasion, he was nevertheless frustrated by the fact that the physics Laureates were speaking at the same time and he desperately wanted to attend the lectures given by Glashow, Salam and Weinberg.[106]

The Cormacks were enjoying breakfast the next morning when Margaret came dashing in and announced there was a dead ringer for 'Dr Who' out in the lobby.[107] The family were enthusiastic fans of the long-running BBC television show by the same name. Jean and Robert raced out to the lobby to take a look and, on their return to the dining room, Robert blurted out, "He doesn't look like Dr Who, that *is* Dr Who!"

They could not remember his real name so, just to be certain, they approached the front desk and asked the duty clerk to page Dr Who. The clerk, whose English was excellent but whose cultural knowledge did not extend to popular British television, took a while to comprehend and carry out the odd request. By then Tom Baker, who was covering the Nobel ceremonies for the BBC, had left the hotel. The two younger Cormacks, who had been watching Baker's response from behind a pillar in the lobby, had disappeared, and were scouring Stockholm for Jelly Babies, Dr Who's favourite sweets, which he ate on the show.

A meeting between Baker and the Cormacks was soon arranged, Jean and Robert having promised to exchange their father's autograph for Baker's. Allan apologised for his children's behaviour but the BBC man insisted that they were having a great time (Figure 7.6). In fact, Tom Baker soon had them in stitches of laughter, describing his visit to California when he bumped into someone who was attending a Star Trek convention. They got chatting and he asked the man what he did. The reply was, "I trek, I trek". Baker then said, "Well, yes, we're all Trekkies at heart, but what is it that you actually do?" He had not realised that his fellow traveller was a full-time Trekkie!

Figure 7.6   Tom Baker, who played the part of Dr Who, entertains Robert Cormack, December 1979.

On Sunday 9 December the Cormacks attended a luncheon given by the Ambassador of the United States, Mr Rodney Kennedy-Minott and his wife at their residence, 2 Nobelgatan. Also attending were the other

American Laureates and their families. Besides the physicists Glashow and Weinberg, there was the chemist Herbert Brown from Indiana and the elderly Theodore Schultz, winner of the economics prize. Margaret was seated at a table with a group of the younger children when the drinks were being served. The children were offered the choice of orange juice or Coca Cola while she had the option of champagne. One of Shelly Glashow's offspring, a self-confident young man no older than ten, said to the waiter, "I'd like a glass of champagne too, please" – and got it![108]

The day they had all been eagerly anticipating – Monday 10 December – finally arrived. After a dress rehearsal in the morning, Lars met the family in the lobby of their hotel at 3:30 that afternoon. Their destination was the Stockholm Concert Hall where, in the Grand Auditorium, the Solemn Festival of the Nobel Foundation was traditionally held. At exactly 4:30 pm, the ten Laureates took their seats on the platform, accompanied by the sound of trumpets. Sure Bergstrom, chairman of the Nobel Foundation, delivered a short speech and welcomed the Laureates. The Stockholm Philharmonic Orchestra played Leonard Bernstein's 'Overture to Candide' after which followed the presentation of the Nobel Prizes for Physics and Chemistry.

Allan's heart soared when the orchestra started playing Beethoven's 'Allegretto Scherzando' from the Symphony Number 8. As the music came to an end, Torgny Greitz strode to the podium to introduce the winners of the Nobel Prize for Physiology or Medicine for 1979, Allan MacLeod Cormack and Godfrey Newbold Hounsfield (Appendix C). Invoking the poetry of Harry Martinson, the Nobel Laureate for Literature in 1974, Greitz described how the mimarobe, the computer guardian, was able to "see through everything as though it were glass". He said they had more than satisfied the provision in Alfred Nobel's will that stipulates a prize winner "shall have conferred the greatest benefit on mankind", and he invited them to receive their insignia from His Majesty, the King. First Allan, followed by Hounsfield, stepped forward to shake King Carl XVI Gustaf's hand and accept their gold medallions (Figure 7.7).

By 5:30 pm the ceremony was over and the Cormacks returned to their hotel to prepare for the banquet, which was hosted by the King and Queen in the City Hall. When they entered the Blue Hall, Barbara was accompanied by the British Ambassador, Sir Jeffrey Petersen. She told him about her experience, over 30 years earlier, of meeting the Queen during her time at Girton College, and of how she had been a little uncertain regarding the correct protocol when introduced to British royalty. In response, the

Figure 7.7   Allan receives the Nobel Prize from the King of Sweden, 10 December 1979.

Ambassador told Barbara that the Swedes were forever asking him if their events like the Nobel Prize presentations and banquet were as grand as those in Britain. Ever the diplomat, he explained to her that he always answered "Oh yes, of course".

The banquet was one of the highlights of their trip to Stockholm, with "lashings of champagne, superb food, and very good company",[109] and a marvellous musical programme. The latter was full of variety, from *Greensleeves* to the *Dance of Zorba the Greek* (the literature Laureate in 1979 was Odysseus Elytis) to Richard Strauss's waltz from *Der Rosenkavalier* (Appendix C). It was also the occasion for Allan to deliver his carefully prepared speech (Figure 7.8):

> Your Majesties, Your Royal Highnesses, Ladies and Gentlemen,
>
> Godfrey Hounsfield has asked me to speak for both of us. We would most respectfully request Your Majesty to convey to the Nobel Foundation and the Nobel Assembly of Karolinska Institutet our intense gratitude for the honour which they have done

us by awarding us the Nobel Prize for medicine and physiology.

There is irony in this award, since neither Hounsfield nor I is a physician. In fact it is not much of an exaggeration to say that what Hounsfield and I know about medicine and physiology could be written on a small prescription form!

While there is irony in the award, there is also hope that even in these days of increasing specialisation there is a unity in the human experience, a unity clearly known to Alfred Nobel by the broad spectrum of his awards. I think that he would have been pleased to know that an engineer and a physicist, each in his own way, have contributed just a little to the advancement of medicine.[110]

This short speech, the only one that evening which successfully adhered to the three-minute time limit,[111] spoke directly of Allan's important character traits: his sense of the occasion, his self-deprecating humour, and his genuine humility.

Figure 7.8   Allan delivers his short speech, on behalf of himself and Godfrey Hounsfield, 10 December 1979.

After a few more hectic days in Stockholm, during which time Allan participated in a round-table TV discussion on 'Science and Man' with the

Laureates in physics and chemistry, the Cormacks left Sweden. On their way home they spent a few days with Godfrey Stafford and his wife Goldie in England. Stafford, who by now had been elected a Fellow of the Royal Society and appointed Director-General of the Rutherford and Appleton Laboratories, had recently accepted a new position as Master of St Cross College at Oxford University.[112] After this brief interlude, during which the two middle-aged Nuclear Nit-Wits and their families reminisced, Allan, Barbara, Margaret, Jean and Robert returned to Winchester for a quiet Christmas. The last three months of 1979 had been an exhilarating period for all of them.

Chapter 8

Citizen of the World

I have always regarded myself as just a physicist, and not a physicist in this or a physicist in that. I enjoy solving problems, wherever I find them.

Allan Cormack[1]

As 1980 dawned, Allan experienced the trappings of a Nobel Prize: instant fame, invitations to speak and an honorary degree. Thirty years previously, after he had abandoned his studies at the Cavendish Laboratory in Cambridge to marry Barbara and return to South Africa, he never imagined that the opportunity might present itself to earn a doctorate. Although his family was aware that he lacked a PhD, it was not a topic on which they ever dwelled. For the three children, who would all go on to earn PhDs in their own fields, their father's lack of a doctorate – the *sine qua non* for all aspiring academics – was considered rather "cool".[2] To Allan, it did not really matter. He had built a successful career at the University of Cape Town and at Tufts University without a PhD. In the spring of 1978 Tufts had awarded the Hosea Ballou Medal to Allan; that accolade was followed two years later on 25 May 1980 with an honorary DSc (Figure 8.1):

Allan Cormack, your pioneering exploration of X-ray techniques has made possible a medical device that has saved countless lives. As an educator, you have wedded research and teaching in the finest tradition of academic scholarship. A proud university honours itself in presenting you with the degree of Doctor of Science, *Honoris Causa*.[3]

In the summer Barbara and Allan travelled to South Africa, their first visit in 23 years. The council of his *alma mater* had voted unanimously to award their distinguished alumnus the Gold Medal of Merit. Allan was the second recipient, the first winner being Christiaan Barnard, the pioneering surgeon who had been awarded the medal after his epic heart transplant at Groote Schuur Hospital in December 1967. On the evening on Monday 14 July 1980, the University of Cape Town fêted its first graduate to win a Nobel Prize. A dinner was held on campus at Smuts Hall, the men's residence named after the late Chancellor of the university, where the guests included Allan's older siblings, Bill and Amy, and their spouses. Also attending were colleagues from the physics department, in addition to Allan's old teachers from Rondebosch Boys' High School, including Mr AA Jayes.[4]

Figure 8.1 Tufts University confers an honorary DSc degree on Allan Cormack, May 1980.

Professor Maurice Kaplan, Deputy Principal of UCT, presented the medal to Allan. In his speech, Kaplan spoke about the dangers of science without humanity:

> Although most of us would accept that science has brought
> great benefits to mankind, it is also realised that various prob-
> lems which face us today, such as the threat of nuclear disaster
> as the arms race continues, have resulted from inventions which
> have followed scientific discoveries.[5]

As a physicist, Allan knew what the implications were. He recalled the
symposium he had attended in the Jameson Hall in August 1945, shortly
after the Allies had dropped the atomic bomb on Hiroshima, and Marthinus
Versveld had spoken about "the extinction of the idea of man".[6] Not only
was the world facing the spectre of nuclear meltdown, South Africa was in
the vice-like grip of apartheid. For Allan and Barbara it was a sobering
time.

Prior to their visit to Cape Town, Allan had been a keynote speaker at
the 25th anniversary meeting of the South African Institute of Physics in
Johannesburg. While in Cape Town he visited the radiology department
at Groote Schuur Hospital, where, with the insight provided by Dr Muir
Grieve, he had first conceived his line integral problem in 1956. While at the
hospital, he saw the new CAT scanner in operation. Also on his itinerary
was a trip to Faure, 15 miles east of Cape Town, where construction of a
cyclotron was under way.[7] Although it was 30 years too late for Allan, the
National Accelerator Centre (now renamed iThemba LABS) would subsequently be recognised as a world class facility, both for the conduct of
particle physics research and for the treatment of cancer patients, using
both protons and neutrons.

Allan and Barbara returned to Cape Town in March 1981, joined by
their daughters Margaret and Jean, for the marriage of Allan's niece, Jo
Read. Sixteen-year-old Robert remained in Winchester, Massachusetts, as
he was preparing for his SAT exams (the scholastic aptitude test for college
entrance). Since he had just passed his driver's licence, Barbara was mildly
concerned! During the two-week visit, Allan gave a short lecture course at
UCT to the mathematicians and physicists on the Radon transform and
its applications. He also delivered a public lecture, on Monday 16 March,
where he spoke on 'Computer Tomography, Past and Future Portent' in the
Robert Leslie Building on University Avenue.[8] When describing the lack
of interest shown in his 1963–64 papers in the *Journal of Applied Physics*,
he recalled the reprint request from Claude Jaccard of Switzerland. His
self-deprecating humour was evident; the audience particularly enjoyed his

anecdote of a CAT scanner being used in the search for a skier buried in the snow.

This was to be Allan's last visit to the land of his birth, although he continued to maintain regular contact with his family and his *alma mater*. In 1996 he was honoured by the country's first democratically elected government when the South African Post Office issued a set of commemorative stamps to celebrate South Africa's Nobel Laureates between 1951 and 1993 (Figure 8.2). The laureates honoured were: Max Theiler (Medicine, 1951); Albert Luthuli (Peace, 1961); Allan Cormack (Medicine, 1979); Aaron Klug (Chemistry, 1982); Desmond Tutu (Peace, 1984); Nadine Gordimer (Literature, 1991); and Nelson Mandela and FW de Klerk (Peace, 1993). The stamps also commemorated the centenary of the death in 1896 of Alfred Nobel (dynamite, which he patented in 1867, had helped to set South Africa on the road to economic prosperity with the development of its mining industry: first diamonds in Kimberley and then gold on the Witwatersrand, both discovered before the end of the 19th century).[9]

Figure 8.2   The South African government commemorates its Nobel Laureates with a set of stamps, November 1996.

Besides the Nobel Prize, the honorary doctorate from Tufts, and the Gold Medal from UCT, there was one other significant award that recognised Allan's contribution to science. On 14 November 1990, Allan joined 19 other scientists in the East Room of the White House where they each received the National Medal of Science from President George Bush:

More and more our nation depends on basic, scientific research to spur economic growth, longer and healthier lives, a
more secure world and indeed a safer environment. If America is to maintain and strengthen our competitive position we
must continue not only to create new technologies but learn
more effectively to translate those technologies into commercial
products.[10]

The medals had been established by President John Kennedy in 1961
to reward a lifetime of work in science and involvement with the science
community. Allan had been nominated by Jean Mayer, President of Tufts,
and the award acknowledged not only his contribution to the development of
the CAT scanner but also his commitment to the teaching of undergraduate
students.[11]

One of the special joys of academia is the friendship that develops between two scholars working on similar topics. Allan's most fruitful collaboration at Tufts was with Eric Todd Quinto, known by his middle name.
Quinto was in the mathematics department and was recognised as an em-
pathetic instructor of undergraduates, who affectionately referred him to
as 'ET' (Steven Spielberg's acclaimed film by the same name had been released in 1982). They knew him to be an excellent teacher, well prepared
and lucid in his explanations.[12] Early in Quinto's career at Tufts, Allan
became concerned that his young colleague was becoming overwhelmed by
responsibilities other than research and so gently reminded him to be more
careful.[13] Quinto must have heeded Allan's advice, for when he was considered for tenure in the fall term of 1983, he was successful, and was also
promoted from assistant to associate professor. While Allan served as a
role model and mentor to Quinto, their relationship was characterised by a
mutual respect between academic peers.[14]

Although Todd Quinto's first collaboration with Allan had been to show
that the paper with Brian Doyle was fatally flawed, their next interaction
was more positively productive. It was an application of the Radon transform to a classical equation, the Darboux partial differential equation, and
Allan told Quinto of a theorem he believed to be true and asked the young
mathematician to prove it.[15] Allan provided a statement of the result, including a constant. Since the constant turned out to be correct, Quinto sur-
mised that his older colleague could have easily proved the theorem himself.
Instead, Allan let Quinto do it and, in this generous way, enabled him to
get started as a mathematician. Their work was completed in August 1979
and a year later their first joint paper was published in the *Transactions of*

the American Mathematical Society.[16]

Allan was also instrumental in introducing Quinto to the leaders in the field of tomography. In February 1980, just two months after Allan had been awarded the Nobel Prize in Sweden, they both attended a colloquium at Oberwolfach in Germany. Located in the beautiful surroundings of the Black Forest, the Mathematical Research Institute was an inspiring place to "do math and enjoy nature".[17] Attendance was by invitation only and their host was Frank Natterer, one of the doyens in the field of mathematics applied to computer tomography. Allan was the keynote speaker and gave a lecture on the early developments in tomography,[18] while Todd Quinto made the acquaintance of leaders such as Gabor Herman, Larry Shepp, Alberto Grünbaum and Bob Marr. A year later, on 11 April 1981, Quinto hosted a conference on the Radon transform on the campus of Tufts University which included a small but select group of experts in the field. In Allan's opinion, the meeting was a great success.[19]

The collaboration between Cormack and Quinto led to two other journal publications, both in the area of radiotherapy.[20] The goal of the therapy was to irradiate the body with ionising radiation, such as X-rays, coming from different directions and to kill the tumour without destroying the healthy tissues. They described the physics of the problem and provided mathematical models of radiation dose planning which were based on the Radon transform. Their close working relationship led Quinto to observe:

> Allan loved to do mathematics, and I anticipated with pleasure hearing about the theorems he was working on and the ideas he had. The mathematical problems he worked on were always important and intriguing. His solutions were elegant – which is a mathematician's way of giving high praise – and he published in excellent journals. Had he wanted a second career, Allan would have been an excellent mathematician (and a great addition to our department!).[21]

One of the major challenges facing Nobel Laureates is what to do for an encore. People ask: does he still possess the creative drive to continue to make seminal contributions to science? Allan was 55 years old when he and his family travelled to Stockholm in December 1979; for many academics, this is the twilight stage of their careers. This was not so in Allan's case. If anything, his productivity actually picked up. Between 1980 and 1995 he published 28 papers (see appendix A) in two broad categories: the mathematics of radiation therapy planning; and the history of computer tomography, with special emphasis on the Radon transform.

‾Allan had first been introduced to radiotherapy at Groote Schuur Hospital in 1956 and, although his 1963 paper in the *Journal of Applied Physics* focused on the imaging problem, he also pointed out how mathematical algorithms could contribute to improved dose planning. In the mid-1980s he returned to this problem with the aim of developing an analytical solution. He was aware that Brahme, Roos and Lax (BRL) had taken the first step in circularly symmetric dose-distributions.[22] Allan then extended the ideas of BRL by studying distributions that were not circularly symmetric, therefore representing the real situation with a patient, using a zero-order approximation.[23] This work was subsequently extended in a collaboration with his son Robert – by now in his final year at Harvard, majoring in physics – when they published the equations for a first-order approximation.[24]

One of the difficulties with using mathematical optimisation to obtain a spherical dose distribution is predicting radiation fields that are negative, a physical impossibility. Allan and Todd Quinto were of course well aware of this problem, and Michael Goitein, while acknowledging the important contributions of BRL and Cormack towards solving the inverse problem in radiation therapy, was nevertheless cautious in his assessment of their approach. He believed the solutions were in their infancy, and that considerable work was still required in order to marry the powerful mathematical approach to a realistic set of clinical criteria.[25] This commentary by Goitein led Allan to respond:

> The questions which Goitein raises about non-negativity and the meaning of 'optimization' are of course valid, but let us remember this: conventional methods of radiotherapy planning are the result of 80 years of careful, painstaking work with an enormous amount of clinical input. By comparison, inverse problems are in their infancy, still mewling and puking in their nurses' arms. Let us nurture these infants for a while (it won't take many journal pages), to see whether they mature into robust adjuncts to radiotherapy planning or whether they die young.[26]

Goitein believed the inverse approach to radiation treatment planning was fundamentally flawed because it presupposed one 'knew' the required 3D dose distribution and it did not account for the trade-off between tumour eradication and normal tissue preservation.[27] However, he acknowledged that both BRL and Allan had introduced the idea of allowing the intensity within a radiation beam to be modulated. The technique has since proved to be enormously valuable.

Throughout the 1980s the Harvard Cyclotron Laboratory continued to lead the world in treating cancer, with 500 patients per annum passing through the facility. While the clinicians and medical physicists at the Massachusetts General Hospital (MGH) would have preferred a higher energy machine located at the hospital in downtown Boston, they enjoyed a symbiotic relationship with the physicists across the Charles River at the lab on the Harvard campus. Then, towards the end of the decade, the Dean of the Faculty of Arts and Science at Harvard, Henry Rosovsky, wrote a letter to Andy Koehler, director of the lab, asking him what his plans were for disengaging from the lab as the university had other plans for the space. With the writing on the wall, Michael Goitein served as the principal investigator on a major grant application submitted to the National Institutes of Health, and in 1994 funds were made available for the construction of a new proton therapy facility.[28]

Herman Suit, head of radiation oncology at the MGH, invited Allan to speak at the ground-breaking ceremony for the Northeast Proton Therapy Center (NPTC), held on the campus of the hospital on 14 September 1995.[29] Allan's topic, 'Behind the Great Leap Forward', provided him with an opportunity to reminisce about the history of the Harvard Cyclotron Laboratory and how it had come to treat cancer patients. Gathered in the audience that day, and publicly acknowledged by Allan, were key players in the history of proton therapy: Bob Wilson, who in 1946 had identified the therapeutic potential of charged particles such as protons; Norman Ramsay, who had been director of the lab during its construction in the late 1940s; Dick Wilson, whose drive and enthusiasm had helped to keep the lab operational for over 40 years; Bill Sweet, chief of neurosurgery at MGH, who had recognised the enormous possibilities of proton therapy; and Andy Koehler, the physicist who supported the medical programme and later served as the lab's director (Figure 8.3).[30]

It would be another five years before the first patient was treated at the NPTC and so the Harvard Cyclotron Laboratory continued to provide this service. Goitein had invited Allan to be a member of the Visiting Committee of the Department of Radiation Oncology at MGH,[31] and so, in addition to his theoretical contributions to radiation dose planning, he was also able to play an active role in the field of radiotherapy (Figure 8.4).

Besides his contributions to radiation therapy, Allan also concentrated on the history of computer tomography, particularly the mathematics of the Radon transform. Despite having discovered Radon's 1917 paper in

Figure 8.3   Andreas Koehler, director of the Harvard Cyclotron Laboratory, and long-time collaborator with Allan Cormack.

the summer of 1971, Allan always had a sneaking suspicion that the so-lution to his line integral problem went back earlier, perhaps to the 19th century. When he and Todd Quinto attended the conference at Oberwol-fach in February 1980, Alberto Grünbaum of the University of California at Berkeley drew their attention to two papers by Dutch mathematicians.[32]

The first of these two publications, which appeared in 1906, was by HBA Bockwinkel and presented his solution to the three-dimensional line integral problem, with application to the propagation of light in a biaxial crystal about a midpoint of oscillation.[33] He stated, without providing an explicit reference, that the problem had been solved by his mentor Hendrik Lorentz, the renowned Dutch physicist and mathematician.

The other paper cited by Grünbaum was published in 1925 by George Uhlenbeck, who also acknowledged Lorentz and generalised the line integral problem to n dimensions.[34] There was an interesting link between Allan and Uhlenbeck: Julian Knipp. Knipp had just finished his PhD on quantum mechanics at Harvard in 1935 when he travelled to Utrecht as a Sheldon Fellow to work with Uhlenbeck. The following year they published a land-mark paper on the emission of gamma radiation during the beta decay of nuclei.[35] Allan subsequently cited their paper in a brief note he published

Figure 8.4   Allan holds a custom-designed beam shaper used in radiotherapy treatment, 1989.

on internal Rayleigh scattering in 1955 in *The Physical Review*[36] and two years later, when he was chairman of physics at Tufts, Knipp offered Allan a job as an assistant professor.

Allan obtained copies of the two Dutch papers and, with his knowledge of Afrikaans, a language he had learned while growing up in South Africa, he was able to translate the relevant passages. Both authors mentioned Lorentz explicitly although neither cited a specific reference. Allan was determined to track down the reference. On the advice of Martin Klein, professor of the history of science at Yale University, he contacted Dr AJ Kox early in 1981.[37] Kox, who worked at the Institute for Theoretical Physics at the University of Amsterdam, was the official editor of Lorentz's papers. He responded in May, indicating that the papers were held in the archives at The Hague, and speculated that the theorem Allan was seeking could well be buried among the huge number of scientific notebooks.[38]

Kox was too busy to search for the theorem and, while Allan had offered to visit The Hague and conduct a systematic search himself, this never happened. Thus, nobody has been able to establish why or even when Lorentz solved the line integral problem because he never published it. This led to an 'in' joke among Allan and his colleagues, to the effect that the Radon transform should really have been called the Lorentz transform, meaning that Lorentz would have had both a transform as well as a transformation named after him.[39] (The Lorentz transformation, one of the cornerstones of the special theory or relativity, converts between measurements of space and time by two different observers, where one observer is in constant motion with respect to the other.)

At about the same time that Allan was attempting to track down the elusive reference by Lorentz, Viktor Ambartsumian, an astronomer based at Yerevan in the Soviet Union, came across Allan's lecture published in *Les Prix Nobel*. He learned that Allan had solved the inverse problem of finding an unknown two-dimensional function from the values of its integrals over lines without knowing about the work of Radon. He was impressed because exactly the same thing had happened to him 45 years previously, in 1936. The mathematical problems were identical, the only difference being that Ambartsumian was working with an astronomical problem of statistics of stellar velocities.[40] His solution was published in the *Monthly Notices of the Royal Astronomical Society* and was based on a problem that had been posed by Arthur Eddington.[41] Two years after publication, someone told Ambartsumian about the work of Johann Radon.

Allan was of course delighted to hear from the legendary Armenian scientist, particularly about the connection with Eddington, his boyhood inspiration. He sent Ambartsumian reprints of all his papers that dealt with the line integral problem, including copies of the articles by Bockwinkel and Uhlenbeck.[42] Over the next 15 years, whenever he gave a lecture on the history of tomography, Allan would always use the work of Ambartsumian as an example of the first practical application of Radon's problem. Whether it was Ron Bracewell studying the radio emission from the sun in Australia, or Aaron Klug uncovering the structure of the human wart virus in Cambridge, or Paul Lauterbur applying gradients to the magnetic field in nuclear magnetic resonance on Long Island, Radon's problem was ubiquitous. As Allan emphasised,

> The variables need not be spatial coordinates at all; they may be physical, chemical, economic, physiological or any number of things. If one of the variables happened to be time, then we

would be able to 'see' some process evolving.[43]

In August 1992, Allan travelled to Vienna where he and other scientists celebrated the 75th anniversary of the Radon transform. Among those attending the conference was Brigitte Bukovics, Radon's daughter, to whom Allan later sent a copy of the book *The Radon Transform and Some of its Applications*, inscribed by the author, Stanley Deans.[44] Allan's articles in the conference proceedings – one about his personal connection with the Radon transform,[45] and the other on mathematical formulas for solving Radon's problem over paraboloids in three-dimensional space[46] – were fittingly two of the last papers that he ever published.

Three years before winning the Nobel Prize, Allan had met a Japanese neurosurgeon, Takayoshi Matsui, at an international symposium hosted by the National Institutes of Health in Maryland. Matsui subsequently requested a photograph of Allan's prototype CAT scanner for a book he was about to publish on computer tomography.[47] This early contact would lead to a series of visits by Allan to Japan, where, as an academic scholar, he was received with a degree of fascination. In January 1981, at the invitation of Matsui, Allan and Barbara travelled to Tokyo where Allan presented the keynote speech to the Japanese Society for Computed Tomography of the Central Nervous System.

The Cormacks arrived a few days before the meeting was due to start and their hosts wined and dined them in excellent French-style restaurants. On the Friday evening Matsui and his colleagues treated them to a traditional Japanese meal. After a seemingly endless variety of dishes, both Allan and Barbara had eaten so well that they decided to fast for the next day and a half.[48] On the Sunday afternoon they were taken to watch Sumo wrestling. During the contest they were offered and ate dried seaweed. Back at the Imperial Hotel Allan was violently ill and, since his important lecture was scheduled for the Monday morning, there was some concern as to whether he would be able to deliver it! Fortunately, he recovered sufficiently to present the lecture which was well received. One of the benefits of such visits was the opportunity for sightseeing. Allan and Barbara particularly enjoyed their trip to Mount Fuji and the picturesque Hakone lake region.[49]

Three years later, in January 1984, Allan received an invitation from Daisuke Yamauchi, President of the Mainichi Newspapers, to contribute an article to the nationallydistributed *Mainichi Shimbun*.[50] The newspaper, which had a circulation of almost five million readers, was hosting a thesis contest with the theme 'Considering the 21st Century – Man and Science'. Yamauchi requested Allan to provide a 2,000-word article to give the readers some 'hints' and offered remuneration of $1,500. Despite the relatively short deadline – just one month – Allan delivered a thought-provoking essay in which he summarised his own scientific philosophy. He was optimistic about a grand synthesis, a convergence between moral and natural philosophies. He identified two profound questions that he thought would remain unanswered in the 21st century. The first was about ultimate cause: 'What preceded the Big Bang?' The second question, no doubt influenced by his reading of EO Wilson, was about biology: 'How is it that we, among all species, are able to separate ourselves sufficiently from our surroundings to be able to think about them?' Allan's ideas have only been exposed to Japanese readers, and so his essay has been reproduced here in its original English (Appendix D).

Allan's next contact with Japan came in March 1988 when he was an invited keynote speaker at the Shueisha Imidas International Seminar, 'The Search for Future Intellect', a rather ambitiously titled theme. He spoke in a session that touched on the technical and moral problems of the 21st century, exploring three related questions to illustrate the complex relationship between technical solutions and moral and even religious principles. The three questions Allan addressed were the world's population, industrial production and pollution. On the last topic he asked:

> Which is better: to produce power from oil and coal and so pollute the atmosphere with CO_2, or to produce more nuclear power and so pollute the earth with radioactive waste? The problem for scientists is to present their conclusions as impartially as they can but this is very difficult to do. Lord Rutherford believed that his investigations of the nucleus would never have any practical applications. How wrong he was![51]

Quoting from a story in the previous day's *Asahi Evening News*, where doctors from Zaire and Egypt claimed to have developed a drug that was 80 percent successful in treating AIDS, Allan lamented the practice of scientists who made unsubstantiated claims of a 'breakthrough'. He challenged his fellow scientists to be far more circumspect when reporting their discoveries, suggesting that they too should re-examine their moral compasses.

Allan's association with Shueisha Incorporated was renewed in December 1991 when Hiroki Goto, editor of *The Weekly Shonen Jump*, a Shueisha publication, visited him at Tufts. The magazine had a primary readership of young people and Goto invited Allan to contribute a short article aimed at inspiring Japanese teenagers to pursue a career in science.[52] The publication that emerged was a collage of photographs, cartoons and brief vignettes, illustrating Allan's life and serving as a self-contained biography.[53] Response to the special series featuring Nobel laureates was such a great success – not only among the young readers, but also among the politicians and government officials and within the scientific and academic communities as well – that Shueisha decided to reprint the articles in book form.[54] The book was published in April 1994, providing further exposure of Allan's love of science and his playful sense of humour to the Japanese people (a translation of the article, with illustrations, has been reproduced in Appendix E).

Lindau, a town on an island at the eastern end of Lake Constance, was for centuries the centre of trade between Switzerland and Bavaria. In 1951, however, two local physicians had the idea of a congress to encourage instead the exchange of scientific ideas between Nobel Laureates and aspiring young scientists. They found an enthusiastic supporter and patron in Count Lennart Bernadotte of Wisborg, from the nearby island of Mainau. Bernadotte's great-grandfather was the Swedish King Oscar II who had presented the first Nobel Prizes in 1901.[55] Thus were born the triennial meetings of Nobel Laureates in Lindau: between 20 and 30 prize winners in one of the three natural science disciplines – physics, chemistry, and physiology or medicine – are joined by about 500 students and young researchers from around the world. For one week, either at the end of June or the beginning of July, there are plenary presentations by Laureates, animated round-table discussions, opportunities for small-group interactions between students and a designated Laureate, and lively social events.

Allan was invited to his first Lindau meeting in 1984 and returned four more times (once every three years), attending his last meeting in the summer of 1996. At the first meeting there were 23 Laureates from Europe and the USA, all of whom had won the prize for physiology or medicine.[56] For Allan it was an opportunity to meet and get to know scientists like Sir John Eccles, who had shared the Nobel Prize in 1963 with Alan Hodgkin and Andrew Huxley for their discovery of the ionic mechanisms of nerve cell membrane, Maurice Wilkins, who had shared his prize with James Watson and Francis Crick for their discovery of the structure of DNA, and Rosalyn Yalow, the grande dame of radioimmunoassay techniques and 1977 Laureate, who had been an outspoken advocate of William Oldendorf in 1979.[57] Allan's plenary presentation on 'Applications of the Principles of CAT Scanning in Science and Medicine' was well received. Heand Barbara enjoyed the occasion, and particularly the opportunity to interact with the young scientists. A group of medical students from the University of Hamburg 'adopted' the American couple and Allan was soon referring to them as his "Hamburgers".[58] He maintained contact with them for several years thereafter, thus fulfilling one of the goals of the meeting.

Although Allan would have enjoyed the chance to attend the physics meetings at Lindau, he nevertheless had ample opportunity to explore the boundaries of science with his fellow medical Laureates. At the 1987 meeting he and Eccles, who by then was aged 84, discussed the latter's recent publication on the application of quantum physics to neuroscience.[59] Eccles hypothesized that mental activity led to neural events in the brain that were analogous to the probability fields of quantum mechanics. The stimulation of interactions such as this meant that Allan always greatly looked forward to these triennial gatherings.

For the meeting in 1990, Count Lennart was succeeded as patron by his wife, Countess Sonja, who by now had come to regard Allan and Barbara as old friends.[60] The final event of the week always consisted of a boat trip from Lindau to Schloss Mainau for a farewell luncheon, hosted by the Count and Countess, with fine food and wine. Allan spoke on behalf of all the guests when he thanked the Bernadottes and their many helpers for their hospitality and organization of another successful week. In 1996, during the return trip from Mainau, a tremendous thunderstorm caught up with their boat as they neared the two lighthouses that stood guard at the entrance to the harbour at Lindau.[61] Some foolhardy sailors ignored the flashing red danger signals around the lake, and Allan saw the mainsails of two boats being blown to shreds while others capsized. That evening a crew member of one of the boats was washed up, drowned in the little harbour outside their hotel. It was a sobering end to an eventful week.

The Nobel Foundation celebrated the ninetieth anniversary of the award of the first prizes in 1991 by inviting all living prize winners to Stockholm. Prior to their departure, a historian friend of Allan's who had a hobby of keeping track of European royal families, reigning or not, came across an announcement in *The Times*: Mr Robert Cormack had been appointed the British ambassador to Sweden. On a whim, Allan looked him up in *Who's Who* and discovered that Robert Cormack's family came from the Orkney Islands, across the Pentland Firth from Allan's own forbears in Caithness (Figure 1.1). Allan subsequently wrote to him, suggesting that they might get together in Stockholm, and received a cordial reply.[62]

As had happened twelve years earlier, when Allan and Barbara arrived in Sweden they discovered that a special programme had been arranged for them and they were accorded VIP treatment. One of the first people they bumped into was Allan's old friend and erstwhile climbing companion, Aaron Klug, now Sir Aaron. Allan told him that in 1985 he had given a lecture in California during which he mentioned Klug's work and its relationship to his own research. Afterwards a man who had been an undergraduate at the University of the Witwatersrand in Johannesburg in the mid-1940s came up and spoke to him. He told Allan that Klug was the only one in their study group who could read Marx in the original German. Klug, however, dismissed this story as apocryphal![63]

On Saturday 7 December there was a symposium held at the Radiumhemmet at the Karolinska Hospital. Allan gave two lectures in the morning, touching on some of the earliest ideas behind tomography, and this was followed by a workshop in the afternoon. Although the audience was small, just 25, it did include Allan's earlier champions, Torgny Greitz and Ulf Rudhe, both by now retired, and a younger group who were doing impressive work. Allan was particularly taken with Anders Brahme – the 'B' in the BRL technique of dose planning – who gave an overview on 'Similarities and differences between computed tomography and radiation therapy planning'.[64]

By Tuesday the 10th, the special day when the prizes were awarded, the Cormacks had not yet bumped into their namesake, the British ambassador. Allan had a word with Aaron Klug who had previously met Robert Cormack and so knew what he looked like, and arranged to bring them together. At four o'clock the next day, Ambassador Cormack's Daimler arrived at the Grand Hotel to take Barbara and Allan to tea at the embassy. Since there were only two Daimlers in Stockholm – the other belonged to the King – they really stood out from the rest of the traffic. Allan later recalled:

> The last time I was in one was in Cape Town in 1946. In anticipation of a royal visit in 1947, the City Council had bought a used one for the Mayor, who was my brother's father-in-law. (For such an august breed of car, I should have used the American phrase 'pre-owned' rather than 'used').[65]

Tea was a very pleasant affair and was also attended by the South African ambassador and his wife, Mr and Mrs Eugene Myburgh. The two Cormacks discovered they had much in common. Robert's father had been in the colonial service in India where he had been responsible for the construction and maintenance of railroads and bridges, while Allan's father had travelled to the colony of Natal where he had dealt with telegraphs and telephones. Both Allan and Robert had studied at Cambridge University, where both met their wives, and each family had two daughters and a son. Their tea was cut short, however, because they all had to prepare for the grand finale: dinner at the royal palace with the King and Queen. One of the highlights of the evening was a concert by the Stockholm Symphony Orchestra, conducted by Solti where the soloist was Dame Kiri Te Kanawa. The musical programme included four pieces by Mozart followed by Brahms' symphony number 1 in C minor, and the orchestra excelled itself.

After ten hectic days, Allan and Barbara returned to the USA with a dozen other prize winners. Among the farewells at Newark and Boston airports could be heard "See you again in 2001", in anticipation of the centenary celebrations. Alas, for Allan, this was not to be.

* * * * * *

In November 1895 Wilhelm Röntgen performed an experiment in which invisible cathode rays, generated by electrostatic discharges from an evacuated glass tube, caused a cardboard screen painted with barium platinocyanide to fluoresce.[66] Röntgen called them X-rays, using the mathematical designation for something unknown. By chance, he stepped into the line of X-rays to test the stopping power of lead and was startled to see an image of his own skeleton shimmering on the fluorescent screen. His groundbreaking article was published just a month later,[67] and the era of medical imaging was under way.

The centenary of Röntgen's discovery was celebrated all over the world in 1995. Since computer tomography was a direct descendant of X-rays, Allan was in demand as a public speaker that year. His old friend Zhang-Hee Cho, whom he had first encountered at the meeting at Brookhaven in 1974, invited him to Seoul, South Korea in May.[68] ZH, as he was known, had been a professor of electrical engineering at the University of California in Los Angeles for many years before returning to his native Korea. He was an acknowledged pioneer in computer tomography, particularly 3-D reconstruction algorithms, as well as magnetic resonance imaging.

Although the Röntgen centenary was the nominal reason for Allan's visit, ZH had also asked him to "make some propaganda" for a 250 MeV proton accelerator for radiotherapy at two hospitals. Allan was fascinated at the extent of technological advance within these hospitals that were run by the philanthropic arms of two industrial giants, Samsung and Hyundai. Digital images from the CAT, MRI and PET scanners (the last-mentioned having its own cyclotron for isotope production), were available, anywhere in the hospital, at the touch of a button.

Allan had visited Seoul once before, in 1982, and was amazed at the rapid development that had occurred in the intervening period. Whole tracts of large modern office buildings existed where 13 years previously there had simply been nothing. The block on which his hotel was situated took 35 minutes to walk around, and overlooked a street that was 14 lanes wide (seven one way and seven the other). While the rapid development had undoubtedly been in preparation for the Olympic Games of 1988, Allan knew enough history to recognise that Seoul had grown from a town of 200,000 in ruins at the end of the Korean War in 1953, to a city of 12.5 million in 1995 – a growth rate of 10 percent per annum compounded for 42 years.

One of ZH's young colleagues took Allan sight seeing and was probably under orders to buy his guest some souvenirs. Despite Allan's protestations that he "liked to travel light", he found himself stuck with a straw hat and three pipes. The pipes, with their thin bamboo stems and small bronze bowls, reminded Allan of turn-of-the-century illustrations of opium dens he had seen in a Sherlock Holmes book. Later, in the hotel elevator, some Australian visitors asked about the pipes and Allan, always mischievous, replied, "Now my wife and daughter and I can smoke our opium the way it ought to be smoked". The horrified looks and shocked gasps suggested the guests were extremely relieved when the 'dope fiend' stepped out of the elevator!

There were three other occasions in 1995 when Allan participated in Röntgen celebrations. On 16 October he spoke at the opening of yet another accelerator – the Advanced Photon Source – at the Argonne National Laboratory in Chicago.[69] The accelerator consisted of a 7 GeV electron synchrotron and was at the time the strongest source of X-rays in the world. Allan was interested to learn that incorporated in the straight sections of the accelerator were a series of 'wigglers' in which alternating magnetic fields caused periodic acceleration of the electrons and thus produced photons. He reminisced how Röntgen's discovery of X-rays in 1895 was followed, in rapid succession, by Henri Becquerel's detection of radioactivity in 1896, JJ Thomson's discovery of the electron in 1897, Max Planck's theory of blackbody radiation in 1901 and Albert Einstein's breakthroughs on relativity

100 Jahre Röntgenstrahlen
1895–1995

Figure 8.5 The centenary of Wilhelm Röntgen's discovery of X-rays is celebrated in November, 1995.

and photons in 1905. All their contributions presaged the development, in 1925, of the modern quantum theory, a theory which continues to surprise and baffle, even those scientists with sophisticated physical intuition.[70]

The highlight of the year was the symposium that took place at the University of Würzburg on 8 November, the precise centenary of Röntgen's discovery of X-rays (Figure 8.5). Professor Dr Axel Haase, chairman of the Würzburg Medical Physics Society, invited Allan to participate in the celebrations that took place in the Neubaukirche.[71] Allan was the only person who spoke in English and, although his knowledge of German was rather rudimentary, he managed to follow the gist of the other speakers' remarks. The one exception was Klaus von Klitzing who spoke so rapidly that even Allan's German-speaking hosts had difficulty understanding him. Allan discovered at the evening feast that von Klitzing was an entertaining dinner-time companion, at equal speed in both German and English![72]

The final event of that auspicious year took place on 21 November, when Allan spoke at the Smithsonian Museum in Washington, DC. There was at the time a distinctly negative attitude towards science among the politicians. As a result, a series of evening lectures aimed at fostering a more positive scientific agenda amongst the denizens of Capital Hill were arranged. Since he had donated his prototype CAT scanner to the Smithsonian some years previously,[73] Allan arranged to have it on display during his lecture. Over 100 people showed up for the event. There were lots of questions, and there would have been more had the chairman not cut them off. Another highlight was the presence of David Hennage who, 32 years previously while he was an undergraduate at Tufts, had gathered the data with the prototype scanner and then written the FORTRAN code that implemented Allan's algorithm for solving his line integral problem. It was a fitting end to the Röntgen year.

Chapter 9

At Home in Massachusetts

It has frequently been a surprise to my listeners that we have here in Massachusetts a nuclear plant which has been pouring forth energy for 29 trouble-free years. Long live Rowe!

Allan Cormack[1]

Among the pleasures of academia is the opportunity to meet with old students. In addition to David Hennage, there were four other Tufts alumni who had a special relationship with Allan: Rick Hauck, Ross Humphreys, Bruce Bernstein and Matt Howard.

Frederick Hauck, known to his friends and family as Rick, came to Tufts in the fall of 1958 from Washington, DC, where his father was in the US military. He earned a degree in physics in 1962 on a Reserve Officer Training Corps (ROTC) scholarship and was then commissioned as an officer in the US Navy. Following further studies in mathematics and nuclear engineering, Hauck received his pilot wings in 1968 and was subsequently deployed to the Western Pacific, serving aboard an aircraft carrier, the USS Coral Sea. He would go on to fly over 100 combat missions during the Vietnam War and receive the Distinguished Flying Cross.[2]

When Hauck was selected as a candidate astronaut by NASA in January 1978, Allan immediately wrote him a letter of congratulations. Two years later, after Allan had won his prize, it was his old student's turn to reciprocate, and he sent a laudatory letter of his own.[3] Rick Hauck would have to wait two more years before flying in space as the pilot of the Space Shuttle on 18 June 1983. His final flight was as the spacecraft commander aboard Discovery on 29 September 1988, NASA's first flight since the

Challenger accident. He retired from active military duty two years later.
William Ross Humphreys graduated with a liberal arts degree from
Tufts in 1971, having struggled through introductory physics in his freshman year. In the fall of 1980 his wife, Susan Lowell, had just given birth
to their first child when she was hospitalised with symptoms of blackouts,
seizures and delirium.[4] The clinicians had great difficulty in establishing a
diagnosis – a psychiatrist even thought she may be suffering from post par-
tum depression – and so she was sent for a CAT scan. While his wife was
being scanned, Humphreys waited in the lobby and began reading an old
copy of *Science* in which Allan's award of the Nobel Prize was described.[5]

The CAT scan image of Susie's brain revealed a meningioma which
was subsequently treated and she made a full recovery. Ross and Susan
Humphreys were so grateful that they donated $1,000 to Tufts University
(in spite of a sizeable balance still outstanding on their student loans), with
instructions for the funds to be used at Allan's discretion.[6] Allan wrote and
thanked Ross and Susan, telling them that their generous gift would go towards a teaching assistantship fund in the physics department. He also
described some of the radiotherapy work at the Harvard Cyclotron Laboratory, and mentioned that the doctors were using protons in the treatment of
meningiomas.[7] Ross Humphreys subsequently went on to establish a successful career in Arizona as a photographer, publisher and ranch owner,
while Susan Lowell became a widely read author of children's books, including *The Three Little Javelinas*, which has sold more than half a million
copies.[8]

Bruce Bernstein graduated from Tufts in 1973 with a degree in applied
physics, *magna cum laude*. He then earned a law degree from Georgetown
University and also served five years as an examiner at the US Patent
and Trademark Office in Washington, DC. He subsequently joined a firm
of patent attorneys, Sandler and Greenblum, and within a few years was
made a partner.[9]

In the summer of 1983 Bernstein visited Tufts for his tenth reunion and
dropped in at Robinson Hall to have a word with his old mentor. The topic
of their conversation was the law suits that EMI had brought against Ohio
Nuclear, Picker, GE and Pfizer, arguing that these companies had infringed
on Hounsfield's US patents. Allan told Bruce about the letter that EMI
had sent to Denis Rutovitz and subsequently sent him a copy.[10] Over the
years Bernstein learned to speak Japanese – scientists from Japan secured
many US patents in the 1980s and 1990s – and he became a managing
partner in the business, which later changed its name to Greenblum and

Bernstein. He went on to serve as President of the Association of Patent Law Firms in 2000 and continues to run a thriving practice.

Matthew Howard III, known as Matt, enrolled at Tufts in 1977. With an undergraduate major in biology, he was a 'pre-med' student, intending to pursue a career in medicine. He showed an aptitude for physics in his introductory course and one of Allan's colleagues persuaded him to consider a double major. Matt signed on, making one of his best decisions:

> The course work was interesting, the upper level classes were small, and the faculty members were uniformly interested in helping students. Allan Cormack's class in medical physics was particularly fascinating because of my future goals. The fundamental knowledge of physics and mathematics that I gained at Tufts has been the basic foundation of all my subsequent work: inventing medical devices, and studying how information is represented in the brain.[11]

When Allan won the Nobel Prize in Matt Howard's third year at Tufts, it made anything seem possible. After graduating in 1981, he entered medical school at the University of Virginia in Charlottesville. When Howard observed neurosurgery procedures during his training he was shocked to discover how crudely the surgeons used the CAT scan images: they would look at the pictures hanging from an X-ray viewing box and then, as far as he could tell, 'guess' how to translate the 3-D information to the patient's actual brain. Matt was certain there had to be a more precise way of 'harnessing' the information and utilising it for precisely controlled surgical interventions in the brain.

While still a medical student at Virginia, Howard conceived of his first invention. He felt the answer to the problem facing neurosurgeons lay in using the precise tomographic information from the CAT scan to focus non-ionising energy at specific sites within the brain.[12] The key concept that evolved was to employ a magnetic field to 'steer' a small metallic object through the brain into the desired target location, and then to heat the object inductively with radiofrequency signals. With further input from neurosurgery resident Sean Grady and physics professor Rogers Ritter, animal trials were conducted and a first US patent was awarded.[13]

In December 1989 Howard and Ritter travelled to Boston to meet with Todd Quinto and Allan Cormack. They had set up a new company, Applied Magnetic Guidance (AMG), and they proposed retaining the two tomography experts as members of the company's scientific advisory board.[14] It soon became clear, however, that an inexperienced group of neurosurgeons and physicists, all of whom had 'day jobs', were completely incapable of running and directing a start-up company that was trying to bring a complicated product to market. AMG was taken over by a venture capital firm from California, Sanderling Ventures, which renamed the company Stereotaxis Incorporated. Sanderling elected not to retain Allan and Todd as scientific consultants, and so they played no further role in the project.

It would take almost a decade before the first magnetic procedure was carried out on a human brain at Barnes Hospital in St Louis, Missouri on 17 December 1998. Stereotaxis survived the long years of development and is now a thriving company with a range of innovative products.[15] After training as a neurosurgeon at the University of Washington in Seattle, Howard spent a year as a senior registrar at Atkinson Morley's Hospital in London in 1991. Although James Ambrose and Godfrey Hounsfield were by then retired, he was able to perform a long-term follow-up study on subdural haematomas in infants based on the hospital's archive of computer tomographic images.[16] Howard returned to the USA where he has enjoyed a distinguished academic career, and currently serves as professor and chairman of neurosurgery at the University of Iowa in Iowa City. To date he has been awarded more than 20 US patents, an astonishing measure of his creativity, which he attributes to the education he received at Tufts under professors such as Todd Quinto and Allan Cormack.

* * * * * *

When one of the pressurized water reactors at the nuclear power station on Three Mile Island near Harrisburg, Pennsylvania, suffered a partial meltdown in March 1979, it was the worst accident of its kind in American commercial power generating history. Even though there was no loss of life, the accident had serious economic and public relations consequences. Just seven years later, on 26 April 1986, there was a major accident at the nuclear power plant at Chernobyl in the Ukraine. An explosion was followed by radioactive contamination over a geographic area that stretched from the western Soviet Union, in a north-westerly direction, across Northern Europe and as far afield as North America. There were more than 50 immediate deaths while many thousands more people were exposed to cancer-causing radiation. The environmental lobby was galvanized and by the summer of 1988 it had managed to persuade the Supreme Judicial Court of the State of Massachusetts to allow a referendum on the November ballot to decide whether the state's two nuclear power plants should be shut down.[17]

As a physicist, Allan knew full well both the risks and benefits of nuclear power. As a Nobel Laureate he had a high visibility in the local community and was invited to join a coalition called 'Massachusetts Citizens Against the Shutdown Initiative'.[18] He readily signed up, as did his old friend at Harvard, Dick Wilson, Mallinckrodt Professor of Physics. Dick also belonged to another group called 'Scientists and Engineers for Secure Energy' which found itself in direct conflict with the 'Union of Concerned Scientists' over the question of nuclear power and its safety.[19] The battle between all the interest groups was not just about safety, however. As *The Boston Globe* argued:

> Thus the shutdown referendum will be, in effect, a referendum on the greenhouse effect – a referendum on whether Americans are going to take sensible steps to avert its dire consequences or are going to wait until the water from a melting Antarctic icecap is lapping the slopes of Beacon Hill.[20]

While the environmental lobby was making the case for wind generators, solar cells and hydroelectricity, these sources of energy would have been inadequate to replace the capacity of the two nuclear power plants in Massachusetts: Yankee at Rowe and Pilgrim at Plymouth. Between them they contributed one fifth of the state's energy needs and, were they to be shut down, the only viable alternative would have been coal- and oil-fuelled plants. The stakes were high and, with the state's governor Michael Dukakis running for President and avoiding local issues, there was uncertainty which way the vote would go. Allan's own position was perfectly captured in the sticker that he displayed on the back bumper of the family's car: 'Split atoms not wood'. He was an enthusiastic supporter of the initiative to "vote 'no' on question 4".[21]

After election day on Tuesday 8 November 1988, when the votes had been tallied, the referendum was soundly defeated by a margin of 68 percent to 32 percent.[22] Dukakis, meanwhile, was also defeated, with George Bush becoming the 41st President of the United States. Andrew Kadak, Chief Operating Officer of the Yankee Atomic Electric Company, wrote to Allan and personally thanked him for his critical support.[23] Allan knew that it had been difficult to be a supporter of nuclear power in New England, given the anti-nuclear stance of government officials, but he was delighted that Yankee's plant at Rowe, which had been operating for 28 trouble-free years, would continue to pour forth energy for the citizens of Massachusetts for many more years to come.[24]

The 1950s and 1960s had been golden years for scientific research in the United States and this enlightened attitude had been one of the reasons for Allan's decision to emigrate from South Africa. Despite the success of NASA's programmes, which were necessarily expensive, it was the Vietnam War that ultimately led to the promulgation of the Mansfield Amendment in the US Congress. Introduced by Senator Mike Mansfield of Montana, the amendment barred the US Defense Department from using its funds "to carry out any research project or study unless such project or study has a direct and apparent relationship to a specific military function".[25] This philosophy that only research with a clear practical function should be supported was carried over into all fields of scholarly endeavour and by the early 1980s there was recognition in the Congress that America's leadership position in the world of science was slipping. Senator John Glenn of Ohio, a celebrated astronaut and champion of government support for scientific research, took the lead in conducting a survey of American Nobel Prize winners.[26]

Letters requesting data were sent to 54 Americans who had won the prize in physics, chemistry or medicine between 1967 and 1981. Only seven scientists failed to respond. Not too surprisingly, the vast majority of respondents (39 of 47) acknowledged that they had received substantial direct support from the federal government. Of the eight who stated they had received little or no direct government funding, most were either industrially supported or had carried out their research prior to World War II when federal funds for research were not available. Allan Cormack was the one scientist who "started his prize-winning work before coming to the US and continued it here as a hobby".[27]

Allan relished the opportunity to engage with Senator Glenn, explaining that while his inclinations were Republican (Glenn's political party), the matters addressed in the survey transcended party politics. He acknowledged that his own case was probably not typical, but also stated that the research done at Tufts and Harvard on the cyclotron had been heavily dependent on funding from the Office of Naval Research (ONR). Allan was unequivocal in his support for Glenn's efforts to halt the proposed cuts in funds for research and development, commenting to the senator:

> The conduct of science has been increasingly turned over to lawyers and bookkeepers instead of the contract managers in the tradition of the ONR who saw their job as facilitating rather than hindering research. And now we have these budget cuts which could turn the crisis into a national disaster.[28]

While Glenn responded positively to Allan's concerns and committed himself to ensuring that the US retained its pre-eminent position as a world leader in research and development,[29] the reality was that federal funding for all research was significantly curtailed during the Reagan-Bush era, from 1980 to 1992. Despite such setbacks, Allan's own commitment to promoting science remained undiminished. On 12 May 1986 he participated in a symposium hosted by the Board on Mathematical Sciences at the National Academy of Sciences in Washington, DC. The moderator was Isadore Singer, the John D MacArthur Professor of Mathematics at MIT, who warmed the audience up by telling them a joke in the form of a riddle.

The story was about a man who parachuted out of a plane, landed in a tree and, when a passerby approached him, asked, "Say, fellow, where am I?" The fellow looked up, thought for a moment, and then responded, "You are in an oak tree dangling from a parachute, 10-and-a-half feet off the ground". The riddle was, "Who is the fellow?" and the answer was, "He is a mathematician". There were three reasons. First, his answer was concise; second, his answer was accurate; and third, his answer was completely irrelevant![30]

The symposium, however, successfully demonstrated the opposite. Allan spoke about the Radon transform and showed how mathematical ideas that were over 80 years old had contributed not just to medical imaging and radiotherapy planning, but to astronomy, electron microscopy and oceanography as well (Figure 9.1). Sharing the podium were two fellow Nobel Laureates: Herbert Hauptman (Chemistry) who won his prize the previous year in 1985, and Steven Weinberg (Physics) who, like Allan, had travelled to Stockholm in 1979. They too provided graphic illustrations of the role that mathematics had played in their own ground-breaking discoveries, Hauptman on the molecular structure of crystals, and Weinberg on string theory, including particle symmetry. The message was clear: mathematics was a unifying thread that ran through all the physical sciences.

Although Allan clearly enjoyed the stimulation of sharing a stage with his fellow prizewinners, he was equally enthusiastic when the stage was less prominent. Just a few months after the symposium in Washington, he was asked by six-year-old Russell Sebring of Florida what he should do to become a scientist. Allan encouraged young Russell to study all the science courses – physics, chemistry, biology, earth sciences and, above all, mathematics – but warned him that he should be prepared for "hard work and hard thinking".[31]

Ten years later, after Allan had retired, his interest in mathematics

Figure 9.1   Allan's love of mathematics comes through clearly in a lecture, 1985.

and its teaching was as strong as ever. He reminisced with his old friend, John Wanklyn, about the time when James and Bragg were undergraduates together at Cambridge in 1910, when the use of vector notation was still far away in the future, and much of the teaching was terrible. He recalled one of the stories that RW James had told him:

> It was about a course in physical optics given by no less an authority than CTR Wilson. It was so bad that he and Bragg had to read a lot to find out what was going on. The result was that both went on to very distinguished careers in crystallography which is, after all, just classical diffraction theory. When he returned to St John's College after achieving some distinction his old tutor used to claim him as one of his most successful pupils but James said, "The old duffer didn't know much mathematics, and a good deal of what he said was wrong!"[32]

It has been said that a recipient of a Nobel Prize, has "an infinite number of degrees of freedom",[33] and of course there are those winners who take advantage of their new-found fame. Michael Goitein recalled a cocktail party where an unending stream of people approached Allan with wide smiles and effusive talk, people who in all likelihood would have ignored him before he received his prize. Goitein thought that their fawning behaviour "resembled nothing so much as the way inferior animals approach the alpha animal in a pack" and realised that Allan could never again trust the motivation of someone who was being friendly towards him.[34]

Fortunately, Allan Cormack was a modest man who remained unaffected by the recognition that the prize conferred on him. When he spoke at the Alberta Heritage Foundation in Edmonton in the Fall of 1981 as the guest of Frank Jackson, one of his old climbing companions from Cape Town,[35] he acknowledged that the award of his prize came down to a great deal of luck and being in the right place at the right time.[36] He encouraged his audience to follow their hunches, even if the work was speculative and they could not gain the necessary financial support from funding agencies. He freely admitted that his most recent grant application, to the National Science Foundation in Washington, DC, had been turned down. Despite his lack of success in grant-writing, Allan was scrupulously honest with his research funds, returning any excess money to Tufts when his out-of-pocket expenses were less than his travel allowance.[37] A further measure of the degree to which Allan remained unaffected by fame was his irrepressible and, at times, acerbic sense of humour.

Figure 9.2   Victor Danilov with 1979 American Nobel Laureates (L to R): Herbert Brown, Allan Cormack, Steven Weinberg and Sheldon Glashow, April 1980.

In April 1980, Allan and Barbara were invited to Chicago by Victor Danilov, President of the Museum of Science and Industry, for the induction of the 1979 American Laureates into the 'Nobel Hall of Science'.[38] Joining Allan were Herbert Brown, Steven Weinberg and Sheldon Glashow

(Figure 9.2). After the black-tie celebration, during which the four in-
ductees spoke of the paths each had taken to the Nobel Prize, Barbara and
Allan retired to their hotel. Established in 1920, The Drake was a landmark
hotel, located in one of Chicago's most exclusive areas, with commanding
views of Lake Shore Drive and Lake Michigan beyond. Allan, however, was
unimpressed with their accommodation that night:

April 25, 1980

The Manager
The Drake Hotel
140 East Walton Street
Chicago, Illinois 60611

Dear Sir:

You were requested by the Museum of Science and Industry to
provide my wife and me with a 'nice double w/double bed' for
the night of April 23. We were assigned to Room 332.

If Room 332 was a nice double, I hate to think what an
ordinary double must be like. Room 332 was small by the
standards of any single room which I can remember. While the
bed may meet some technical specification as 'double', it was
the smallest double bed I have ever slept in. Furthermore,
only two rather thin pillows were provided, with no back-up
pillows in the closet.

My wife and I had a very uncomfortable stay in Room 332, and
I hope you will have it reclassified as a small single.

Sincerely

A.M. Cormack

A.M. Cormack
Professor of Physics

In 1994, the year in which Allan turned 70, universities in the USA abolished the mandatory retirement age. While administrators were uncertain of the impact this policy would have on their institutions, Stephen Stigler, a distinguished professor of statistics at the University of Chicago, provided an analysis of data from 18th century England, which appeared in *Science*.[39] He reported that of the 20 men to hold professorships of mathematics or astronomy at Cambridge and Oxford at some time during that century, 18 of them had held their chair until death, one was banished for heresy and only one had actually retired voluntarily.

Allan was delighted to read the report, contacted Stigler immediately and asked to know the heresy for which five percent of his sample had been banished, suggesting that, "Perhaps this would inspire me to find a modern equivalent, say a delicious piece of Political Incorrectness, which would get me booted out!"[40] Stigler responded with the information that the banished professor was William Whiston (1667–1752), who had succeeded Isaac Newton as holder of the Lucasian Chair of Mathematics at Cambridge in 1701. In 1710 he had been "deprived of his chair and driven from the university", his crime being anti-Trinitarianism, a flaw he was incautious enough to reveal in his public writing.[41] Stigler suspected that at Tufts, as at Chicago, it would have taken a far more serious crime to achieve the same end in 1994!

When Allan studied at Cambridge University in the late 1940s, he had visited his mother's siblings Betty and John in Wick and they were instrumental in arousing his genealogical interests. For the next 50 years, on and off, he pursued this interest. When his aunt and uncle died in the early 1970s, without writing down some of the family stories he had begged them to record, Allan thought that his connections to any relatives living in Scotland had been irrevocably lost. However, after he won his prize in 1979 his name was in newspapers and he began to hear from Cormacks located all over the world, but, with one exception, none was related. The exception was Isobel Sinclair (née Cormack), a second cousin still living in Wick, who reminded Allan that they had first met back in 1938 at the home of their grand-uncle Alex.[42]

Happily, Isobel was also instrumental in locating Allan's first cousin, Allan MacLeod, whom he had never met because of the family feud. In 1981 Allan and Barbara visited the MacLeods in Giffnock, just south of Glasgow, where they both taught at the local high school. The two Allans later exchanged family trees and continued to correspond. Allan Cormack subsequently joined the New England Historical Genealogical Society and through them discovered the International Genealogical Index that was created and operated by the Mormon Church. He also established that one could hire people to do genealogical research in other parts of the world, although this of took away much of the pleasure of self-discovery and the feeling of doing detective work. So it was in 1992 that Allan began corresponding with Dr Ewen Collins of Kircaldy, in Fife, who helped him to assemble a document of the family's history.

After retiring from Tufts in 1994, Allan was able to spend more time investigating his family history. Since his parents were first cousins, he had only six great-grandparents (Figure 1.2), and he was especially pleased to have photographs of three of them (Figure 1.3). With the assistance of his sister Amy, and her husband Len, he assembled a photo album of their Scottish ancestors, including pictures of Pulteneytown, where all the Cormacks and MacLeods had lived.[43] One of his motivations for documenting the family's history was to provide his own children, two of whom had been born in South Africa, with some idea of their own origins.[44] As Allan thought about his own family and how they had been formed by history and, in their own small way, had contributed to it, he was quietly proud of the accomplishments of Margaret, Jean and Robert (Figure 9.3).

Figure 9.3   The Cormack family in Iceland (L to R): Barbara, Allan, Margaret, Robert and Jean, 1986.

Margaret, or Meg as she is known to her friends, graduated with a BA degree in linguistics and Germanic languages from Harvard in the summer of 1974. There followed two years of travelling and studying in Germany and Scotland, after which she enrolled for a postgraduate degree in medieval studies at Yale University in New Haven, Connecticut. As a graduate student she spent a year in Iceland, which became the setting for her doctoral dissertation, later published in book form as *The Saints in Iceland*. After working as a research assistant at Harvard University for a few years, Meg secured her first academic appointment as a visiting lecturer at Smith College in Northampton, Massachusetts, first teaching history and then religious studies. Since Allan was somewhat impatient with his elder daughter who, now in her early 40s, did not yet have a permanent job,[45] he was cautiously optimistic when, in the summer of 1995, she accepted a position as Associate Professor of Philosophy and Religious Studies at the College of Charleston in South Carolina. Meg, who spends each summer in Reykjavik, pursuing her scholarly interests in Iceland and saints, continues to teach at the College of Charleston.

Jean earned a BA degree from Tufts University, majoring in music, where her instrument was the violin, graduating in the summer of 1980, at the same ceremony at which her father was awarded an honorary DSc. In the mid-1980s Jeanie, as she is known to her family, enrolled at Iowa State University in Ames, Iowa, earning a master's degree in plant breeding and cytogenetics in 1987. The following seven years she worked as a geneticist and statistician for the Garst Seed Company in Slater, Iowa. Jeanie next took a position as corn-testing manager for DEKALB Genetics, based in DeKalb, Illinois, and also managed to complete her PhD in plant breeding from Iowa State University in 1997, with her thesis concentrating on a stability analysis of single-cross hybrids of maize. Since the beginning of 2003 she has worked as a biostatistician at Brown University in Providence, Rhode Island, working on cancer-screening trials for the American College of Radiology Imaging Network. With experience in the analysis of various medical imaging modalities – including digital X-rays, magnetic resonance imaging, ultrasound and computer tomography – Jeanie's most recent work has involved analysis of CT contours of head and neck tumours.[46]

Although chemistry was one of Robert's favourite subjects at high school (French was most definitely his least favourite), he chose to major in physics when he entered Harvard in the Fall of 1982. Having met Shelly Glashow in Stockholm in December 1979, Robert requested the Nobel Laureate as his advisor. Unbeknown to Glashow, who enjoyed working with undergraduates, the secretary of the physics department had decided to relieve him of that responsibility and told the young Cormack, "Oh no, Professor Glashow doesn't take advisees".[47] Robert settled for his father's old colleague and then chairman of the department, Dick Wilson. After graduation he spent a year at CERN in Geneva, Switzerland, doing high energy physics where, ironically, his fluency in French would come in handy. On his return to the USA in the summer of 1987, Robert was recruited by Larry Sulak of the physics department at Boston University to pursue a doctorate. During the next six years he spent time in Italy where he conducted research at a large underground detector in a cave beneath the Gran Sasso, about two hours' drive east of Rome. The detector, known as the Monopole and Cosmic Ray Observatory (MACRO), enabled Robert to gather the experimental data for his PhD which he was awarded in January 1994.

Faced with the prospect of being just one of over 100 co-authors on his journal papers, and much to the consternation of his father, Robert switched from high energy physics to the physics of radiation therapy: he accepted a post-doctoral fellowship in medical physics at Harvard Medical School. During this period, Robert re-visited the paper that he and Allan had published ten years previously and established that it was possible to create concave distributions for the treatment beam, thus protecting sensitive structures such as the spinal cord.[48] Robert later married Karin Jean Froom and they have three daughters, Kiersten, Monika and Lilja. Since 2001 he has been an Assistant Professor of Radiation Oncology, affiliated with Harvard University and the Brigham and Women's Hospital in Boston.

Despite having smoked cigarettes all his adult life, Allan had enjoyed relatively good health. That changed in September 1990 when he awoke at 2 o'clock in the morning to find his heart beating furiously and erratically. Subsequent tests revealed an atrial fibrillation that was triggered by an excess of cigarettes and alcohol. Although the condition was described by his physician as "not life-threatening, just a nuisance", Allan stopped smoking, limited his alcohol intake, and commenced daily medication of Cuminin, a blood-thinning agent. Three days in hospital were followed by three weeks of enforced bed rest at home, during which time he was so bored he was even reduced to watching a little daytime TV, although, "it was a *very* little, because daytime TV is so bloody awful!"[49] It was a great relief for Allan when he was able to return to work.

In the summer of 1994, while the World Cup soccer tournament was being hosted by the USA, Allan and Barbara flew to the UK, landing at Heathrow Airport in London. They were on their way up to Caithness where Allan was planning to continue his quest for the family history. He was shuffling along in a queue with his bags on the floor, pushing the bags ahead of him, when he tripped over them.[50] Allan ruptured the quadriceps tendons in both legs and, as a result of the blood-thinning medication he was on, he bled profusely. After emergency surgery to repair the tendons, and two weeks of recuperation in a London hospital, during which period Barbara stayed with Godfrey and Goldie Stafford in Oxford,[51] Robert managed to arrange a return flight to Boston. With both legs in casts, his knees fully extended, Allan had to fly first class. After some months of rehabilitation he was able to return to work, but whenever he flew thereafter, he was obliged to travel first class to ensure that he had enough stretching room for his legs. Since he had always been dissatisfied with the cramped seating of economy class, it gave him great delight to have 'doctor's orders' on which to base requests for a comfortable flight.[52]

The Cavendish Laboratory at Cambridge decided to celebrate the centenary of JJ Thomson's discovery of the electron on 21 March 1997. Allan, along with other leading scientists, was invited to join the celebrations. He looked forward to renewing his friendship with colleagues such as Aaron Klug, Godfrey Stafford and others at St John's College, and hearing about the contribution of electrons to chemistry, biology, integrated circuits, tunneling microscopes and holography.[53] However, it was not to be. Just three days before the celebrations were due to begin Allan was diagnosed with a malignancy in his bile-duct. Bitterly disappointed, he sent an e-mail message to Archie Howe, the Cavendish Professor, to inform him that a medical emergency had forced the cancellation of the visit.[54]

At the time of his diagnosis, the doctors gave Allan five years to live. A liver transplant was not an option, nor was any other kind of surgery. They were able to place a stent in the bile duct, which worked for a few months before blocking up. Allan's philosophy was to accept the hand that he had been dealt: he was adamant that there should be no heroic efforts to keep him alive.[55] He maintained regular contact with his old friends, Walter Powrie and John Wanklyn. In early 1998 the BBC broadcast a TV programme about climbing in the Western Cape, and Wanklyn, living in Oxford, recorded the show and sent the videotape to Allan. Having converted the tape from PAL, the British format, into the American VHS format, Allan watched the recording. He subsequently sent an e-mail to his old friend on Sunday evening, 1 March 1998:

> Before commenting on the tape, here's a little relevant geology. The Table Mountain sandstone is a remarkably horizontal series of strata stretching all the way up to the Cederberg. This horizontal structure makes for many so called 'dassie traverses' – so called because they often contain the droppings of dassies –

the coneys of the Bible. The cracks perpendicular to the strata
provide the principal means of vertical ascent, as in the tape
where they surmounted these with a couple of fierce looking
'lay-backs'.

Below the cable station there is a ledge called Fountain Ledge,
and this last 400 feet of Table Mountain must contain some of
the finest rock climbing in the world. I have a picture of Walter
Powrie on the last pitch of a climb called 'Staircase'. If he had
come off at that point, and if he had not been roped, he would
certainly have fallen to his death 400 feet below. However, by
angling the camera just a little it appears as if he would have
fallen into the sea at Clifton – 3,600 feet below!

I was surprised by the comments on non-whites climbing in the
Cape. During the late forties and early fifties there was a very
active group of Cape Coloured climbers with their own club.
At that time there were earnest discussions among members of
the UCT Mountain and Ski Club to see whether we should offer
our fellow climbers the use of the Waaihoek Hut if they desired
it. The decision reached was not to do so, a cowardly decision,
perhaps, but not without force at the time.[56]

At the age of 74, while he was sitting in his study and typing up an
e-mail, Allan's thoughts turned to his climbing heritage ten thousand miles
away in the Western Cape of South Africa, to his beloved Table Mountain, the Waaihoek Hut and the unforgettable Cederberg. Just two months
later, on 7 May 1998, Allan Cormack passed away at home, in Winchester,
Massachusetts.

According to his wishes, Allan's body was cremated and the ashes placed
in an urn in the side of a granite rock face in a quarry near Lake Winnepesaukee in Alton, New Hampshire, where several Seaveys had already been
interred. On Tuesday 30 June, a memorial service was held in the Goddard
Chapel at Tufts. The University Chaplain, the Reverend Scotty McLennan,
in his eulogy, quoted from EO Wilson's *Consilience* that, "All people yearn
to have a purpose larger than themselves". He made mention of Allan's
contributions to science, particularly computer tomography, then finished
with these words:

Yet Allan was also a wonderful teacher, colleague, friend, husband, father and grandfather. He was a modest and unassum-
ing man and enjoyed people.[57]

Appendix A

Allan Cormack's Publications

Stoy, R. H. and Cormack, A. (1945). On the photometric use of plates taken for the determination of stellar parallax, *Monthly Notices of the Royal Astronomical Society*, **105**, 4, pp. 225–236, 1945.

Cormack, A. M. (1947). X-ray powder data of the Parkerite series $Ni_3Bi_2S_2 - Ni_3Pb_2S_2$, *Transactions of the Geological Society of South Africa*, **50**, pp. 17–22.

Cormack, A. M. (1953). Momentum determination from cloud-chamber photographs in axially symmetric magnetic fields, *The Review of Scientific Instruments*, **24**, 3, p. 232.

Cormack, A. M. (1954). Interference of Rayleigh and nuclear Thomson scattering, *The Physical Review*, **94**, 5, pp. 1397–1398.

Cormack, A. M. (1954). Heat generation in the earth by solar neutrinos, *The Physical Review*, **95**, 2, pp. 580–581.

Cormack, A. M. (1954). Resonance scattering of gamma rays by nuclei, *The Physical Review*, **96**, 3, pp. 716–718.

Cormack, A. M. (1955). Neutrinos from the sun, *The Physical Review*, **97**, 1, pp. 137–139.

Cormack, A. M. (1955). Internal Rayleigh scattering, *The Physical Review*, **97**, 4, p. 986.

Cormack, A. M. (1956). Some physical aspects of the corona, *Monthly Notices of the Astronomical Society of South Africa*, **15**, pp. 2–6.

Cormack, A. M. (1957). Fourier transforms in cylindrical co-ordinates, *Acta Crystallographica*, **10**, pp. 354–358.

Palmieri, J. N., Cormack, A. M., Ramsey, N. F. and Wilson, R. (1958). Proton-proton scattering at energies from 46 to 147 MeV, *Annals of Physics*, **5**, 4, pp. 299–341.

Cormack, A. M., Palmieri, J. N., Ramsey, N. F. and Wilson, R. (1959).

Elastic scattering and polarization of protons by Helium at 147 and 66 MeV, *The Physical Review*, **115**, 3, pp. 599–608.

Cormack, A. M. (1959). Semiclassical theory of internal Rayleigh scattering, *The Physical Review*, **115**, 3, pp. 619–623.

Cormack, A. M., Palmieri, J. N., Postma, H., Ramsey, N. F. and Wilson, R. (1960). Elastic $(p - \alpha)$ and $(p - d)$ scattering at 147 and 66 MeV, *Nuclear Forces and the Few Nucleon Problem*, Volume 1, pp. 259–268, Pergamon Press.

Cormack, A. M. (1962). Dead-time losses with pulsed beams, *Nuclear Instruments and Methods*, **15**, pp. 268–272.

Cormack, A. M. (1963). Representation of a function by its line integrals, with some radiological applications, *Journal of Applied Physics*, **34**, 9, pp. 2722–2727.

Cormack, A. M. (1963). Diffraction by atoms distributed at random at lattice sites, *Acta Crystallographica*, **16**, pp. 1051–1055.

Palmieri, J. N., Goloskie, R. and Cormack, A. M. (1963). Normalization of p-p, p-d and p-He differential cross sections measured at the Harvard Cyclotron Laboratory, *Physics Letters*, **6**, 3, p. 289.

Cormack, A. M. (1964). The effects of multiple scattering in small-angle nuclear scattering experiments, *Nuclear Physics*, **52**, pp. 286–300.

Steinberg, D. J., Palmieri, J. N. and Cormack, A. M. (1964). Small-angle scattering of 143 MeV polarized protons, *Nuclear Physics*, **56**, pp. 46–64.

Cormack, A. M. (1964). Representation of a function by its line integrals, with some radiological applications. II, *Journal of Applied Physics*, **35**, 10, pp. 2908–2913.

Widgoff Shapiro, M., Cormack, A. M. and Koehler, A. M. (1965). Measurement of cross sections with neutrons as targets, *The Physical Review*, **138**, 4B, pp. B823–B830.

Ayres, D. S., Cormack, A. M., Greenberg, A. J., Kenney, R. W., Caldwell, D. O., Elings, V. B., Hesse, W. P. and Morrison, R. J. (1968). Precise comparison of π^+ and π^- lifetimes, *Physical Review Letters*, **21**, 4, pp. 261–265.

Schneider, R. J. and Cormack, A. M. (1968). Neutron total cross sections in the energy range 100 to 150 MeV, *Nuclear Physics*, **A119**, pp. 197–208.

Widgoff, M., Cormack, A. M. and Koehler, A. M. (1969). Measurement of cross sections with neutrons as targets. II, *The Physical Review*, **177**, 5, pp. 2016–2022.

Ayres, D. S., Cormack, A. M., Greenberg, A. J., Kenney, R. W., McLaughlin, E. F., Schafer, R. V., Caldwell, D. O., Elings, V. B., Hesse, W. P. and Morrison, R. J. (1969). An improved liquid hydrogen differential Čerenkov counter, *Nuclear Instruments and Methods*, **70**, pp. 13–19.

Ayres, D. S., Cormack, A. M., Greenberg, A. J., Kenney, R. W., Caldwell, D. O., Elings, V. B., Hesse, W. P. and Morrison, R. J. (1969). A determination of the refractive index of liquid deuterium using the Čerenkov effect, *Physica*, **43**, pp. 105–108.

Greenberg, A. J., Ayres, D. S., Cormack, A. M., Kenney, R. W., Caldwell, D. O., Elings, V. B., Hesse, W. P. and Morrison, R. J. (1969). Charged-pion lifetime and a limit on a fundamental length, *Physical Review Letters*, **23**, 21, pp. 1267–1270.

Ayres, D. S., Cormack, A. M., Greenberg, A. J., Kenney, R. W., Caldwell, D. O., Elings, V. B., Hesse, W. P. and Morrison, R. J. (1971). Measurements of the lifetimes of positive and negative pions, *Physical Review D. Particles and Fields*, **3**, 5, pp. 1051–1063.

Cormack, A. M. (1973). Reconstruction of densities from their projections, with applications in radiological physics, *Physics in Medicine and Biology*, **18**, 2, pp. 195–207.

Galvin, J. M., Bjarngard, B. E. and Cormack, A. (1975). *In vivo* determination of attenuation coefficients for radiation therapy, Third Annual Meeting on Applications of Optical Instrumentation in Medicine, *Proceedings of the Society of Photo-Optical Instrumentation Engineers*, **47**, pp. 27–39.

Cormack, A. M. and Koehler, A. M. (1976). Quantitative proton tomography: preliminary experiments, *Physics in Medicine and Biology*, **21**, 4, pp. 560–569.

Cormack, A. M., Koehler, A. M., Brooks, R. and Di Chiro G. (1977). Proton tomography, International Conference on Computer Assisted Tomography in Nontumoral Diseases of the Brain, Spinal Cord and Eye, *Journal of Computer Assisted Tomography*, **1**, p. 266.

Cormack, A. M. and Doyle, B. J. (1977). Algorithms for two-dimensional reconstruction, *Physics in Medicine and Biology*, **22**, 5, pp. 994–997.

Cormack, A. M. (1978). Sampling the Radon transform with beams of finite width, *Physics in Medicine and Biology*, **23**, 6, pp. 1141–1148.

Cormack, A. M. (1980). An exact subdivision of the Radon transform and scanning with a positron ring camera, *Physics in Medicine and Biology*, **25**, 3, pp. 543–544.

Cormack, A. M. (1980). Early two-dimensional reconstruction and recent topics stemming from it, *Les Prix Nobel en 1979*, Nobel Foundation, Stockholm, pp. 171–183. It also appeared in: *Science*, **209**, 4464, pp. 1482–1486; *Medical Physics*, **7**, 4, pp. 277–282; and *Journal of Computer Assisted Tomography*, **4**, 5, pp. 658–664.

Cormack, A. M. (1980). Recollections of my work with computer assisted tomography, *Molecular and Cellular Biochemistry*, **32**, pp. 59–61. This article is reprinted from the November 1979 issue of *Tufts Criterion*.

Cormack, A. M. and Quinto, E. T. (1980). A Radon transform on spheres through the origin in R^n and applications to the Darboux equation, *Transactions of the American Mathematical Society*, **260**, 2, pp. 575–581.

Cormack, A. M. (1980). Algorithms for two-dimensional reconstruction (letter), *Physics in Medicine and Biology*, **25**, 2, p. 372.

Cormack, A. M. (1981). Early tomography and related topics, in *Mathematical Aspects of Computerized Tomography*, edited by GT Herman and F Natterer, Springer-Verlag, New York, pp. 1–6.

Cormack, A. M. (1981). The Radon transform on a family of curves in the plane, *Proceedings of the American Mathematical Society*, **83**, 2, pp. 325–330.

Cormack, A. M. (1981). On the future of the principles of computed tomography, *Toshiba Medical Review*.

Cormack, A. M. (1982). The Radon transform on a family of curves in the plane. II, *Proceedings of the American Mathematical Society*, **86**, 2, pp. 293–298.

Cormack, A. M. (1982). The future of the principles of computed tomography, *Proceedings of the Fourth Philip Morris Science Symposium*, edited by CG Lunsford, Philip Morris, New York, pp. 38–50.

Cormack, A. M. (1983). Computed tomography: history and some recent developments, *Proceedings of the International Conference on Applications of Physics to Medicine and Biology*, edited by G Alberi, Z Bajzer, and P Baxa, World Scientific Publishers, Singapore, pp. 381–390.

Cormack, A. M. (1983). Computed tomography: some history and recent developments, in *Computed Tomography*, edited by LA Shepp, Proceedings of Symposia in Applied Mathematics, American Mathematical Society, Providence, Rhode Island, **27**, pp. 35–42.

Cormack, A. M. (1984). Radon's problem – old and new, *SIAM-AMS Proceedings*, **14**, pp. 33–39.

Cormack, A. M. (1985). Electrostatics revindicated by conducting ellipsoid, *Nature*, **316**, pp. 301–302.

Griffiths, P. A., Singer, I. M., Cormack, A. M., Hauptman, H. A. and Weinberg, S. (1986). Mathematics: the unifying thread in science, *Notices of the American Mathematical Society*, **33**, pp. 716–733.

Cormack, A. M. (1987). Radon's problem for some surfaces in R^n, *Proceedings of the American Mathematical Society*, **99**, 2, pp. 305–312.

Cormack, A. M. (1987). Scanning in medicine and other fields, *Cell Biophysics*, **9**, pp. 151–169.

Cormack, A. M. (1987). A problem in rotation therapy with X rays, *International Journal of Radiation Oncology, Biology, Physics*, **13**, 4, pp. 623–630.

Cormack, A. M. and Cormack, R.A. (1987). A problem in rotation therapy with X-rays: dose distributions with an axis of symmetry, *International Journal of Radiation Oncology, Biology, Physics*, **13**, 12, pp. 1921–1925.

Cormack, A. M. (1988). Early developments of computed tomography and unusual applications today, *Modern Imaging: State of the Art*, Proceedings of the Schering Symposium, edited by T Vogel, Ueberreuter, Berlin.

Cormack, A. M. and Quinto, E. T. (1989). On a problem in radiotherapy: questions of non-negativity, *International Journal of Imaging Systems and Technology*, **1**, pp. 120–124.

Cormack, A. M. (1990). Response to Goitein (letter), *International Journal of Radiation Oncology, Biology, Physics*, **18**, p. 709.

Cormack, A. M. (1990). Computed axial tomography, in *Technology of Our Times*, SPIE Optical Engineering Press, edited by Frederick Su, Bellingham, Washington, pp. 96–101.

Cormack, A. M. and Quinto, E. T. (1990). The mathematics and physics of radiation dose planning using X-rays, *Contemporary Mathematics*, **113**, pp. 41–55.

Cormack, A. M. (1992). 75 years of Radon transform, *Journal of Computer Assisted Tomography*, **16**, 5, p. 673.

Cormack, A. M. (1994). EMI patent litigation in the US (letter), *British Journal of Radiology*, **67**, 795, pp. 316–317.

Cormack, A. M. (1994). My connection with the Radon transform, *Proceedings of the Conference 75 Years of the Radon Transform*, International Press, pp. 32–35.

Cormack, A. M. (1994). A paraboloidal Radon transform, *Proceedings of the Conference 75 Years of the Radon Transform*, International Press, pp. 105–109.

Cormack, A. M. (1995). Some early radiotherapy optimization work, *International Journal of Imaging Systems and Technology*, **6**, 1, pp. 2–5.

Appendix B

Early Two-Dimensional Reconstruction and Recent Topics Stemming from It

On 8 December 1979 Allan MacLeod Cormack delivered his Nobel Lecture in Stockholm, Sweden. The lecture appeared in *Les Prix Nobel en 1979* and in three separate journals: *Science*, 209(4464): 1482-1486, 1980; *Medical Physics*, 7(4): 277-282, 1980; and *Journal of Computer Assisted Tomography*, 4(5): 658-664, 1980. It is reproduced here with the permission of the Nobel Foundation.

In 1955 I was a Lecturer in Physics at the University of Cape Town when the Hospital Physicist at the Groote Schuur Hospital resigned. South African law required that a properly qualified physicist supervise the use of any radioactive isotopes, and since I was the only nuclear physicist in Cape Town, I was asked to spend $1\frac{1}{2}$ days a week at the hospital attending to the use of isotopes, and I did so for the first half of 1956. I was placed in the Radiology Department under Dr. J. Muir Grieve, and in the course of my work I observed the planning of radiotherapy treatments. A girl would superpose isodose charts and come up with isodose contours which the physician would then examine and adjust, and the process would be repeated until a satisfactory dose-distribution was found. The isodose charts were for homogeneous materials, and it occurred to me that since the human body is quite inhomogeneous these results would be quite distorted by the inhomogeneities – a fact that physicians were, of course, well aware of. It occurred to me that in order to improve treatment planning one had to know the distribution of the attenuation coefficient of tissues in the body, and that this distribution had to be found by measurements made external to the body. It soon occurred to me that this information would be useful for diagnostic purposes and would constitute a tomogram or series of tomograms, though I did not learn the word 'tomogram' for many years.

At that time the exponential attenuation of X- and gamma-rays had been known and used for over sixty years with parallel sided homogeneous slabs of material. I assumed that the generalization to inhomogeneous materials had been made in those sixty years, but a search of the pertinent literature did not reveal that it had been done, so I was forced to look at the problem *ab initio*. It was immediately evident that the problem was a mathematical one which can be seen from Figure B.1. If a fine beam of gamma-rays of intensity I_0, is incident on the body and the emerging intensity is I, then the measurable quantity $g = ln(I_0/I) = \int_L f ds$, where f is the variable absorption coefficient along the line L. Hence if f is a function in two dimensions, and g is known for all lines intersecting the body, the question is: "Can f be determined if g is known?". Again this seemed like a problem which would have been solved before, probably in the 19th century, but again a literature search and enquiries of mathematicians provided no information about it. Fourteen years would elapse before I learned that Radon had solved this problem in 1917. Again I had to tackle the problem from the beginning. The solution is easy for objects with circular symmetry for which $f = f(r)$, r being the radius. One has Abel's equation to solve, and its solution

$$f(r) = -\frac{d}{dr}\left[\frac{r}{\pi}\int_r^\infty \frac{g(s)ds}{s(s^2 - r^2)^{\frac{1}{2}}}\right] = -\frac{dI(r)}{dr} \tag{B.1}$$

has been known since 1825. In 1957 I did an experiment in Cape Town on a circularly symmetrical sample consisting of a cylinder of aluminium surrounded by an annulus of wood. The results are shown in Figure B.2. Here $I(r)$ is plotted against r and the constant slopes indicate the constant values of the absorption coefficient in wood and aluminium. Even this simple result proved to have some predictive value for it will be seen that the three points nearest the origin lie on a line of a slightly different slope from the other points in the aluminium. Subsequent enquiry in the machine shop revealed that the aluminium cylinder contained an inner peg of slightly lower absorption coefficient than the rest of the cylinder.

Further work occurred intermittently over the next six years. Using Fourier expansions of f and g, I obtained equations like Abel's integral equation but with more complicated kernels, and I obtained results like

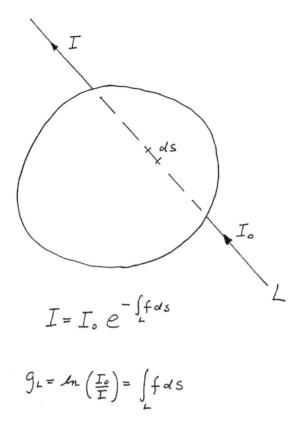

$$I = I_o \, e^{-\int_L f \, ds}$$

$$g_L = \ln\left(\frac{I_o}{I}\right) = \int_L f \, ds$$

Figure B.1　Beam of gamma rays of initial intensity I_0 passing through body along line L and emerging with intensity I.

equation B.1, but with more complicated integrands. However, they were also integrals from r to ∞, a point which I shall return to. These integrals were not too good for data containing noise, so alternative expansions were developed which dealt with noisy data satisfactorily. By 1963 I was ready to do an experiment on a phantom without circular symmetry with the apparatus shown in Figure B.3. The two cylinders are two collimators which contain the source and the detector, and which permit a 5 mm beam of ^{60}Co gamma-rays to pass through the phantom which is the disc in between them. At that time I was approached by an undergraduate, David Hennage, who wanted to know whether I had a numerical problem for him to work on so that he could learn FORTRAN and learn how to use a

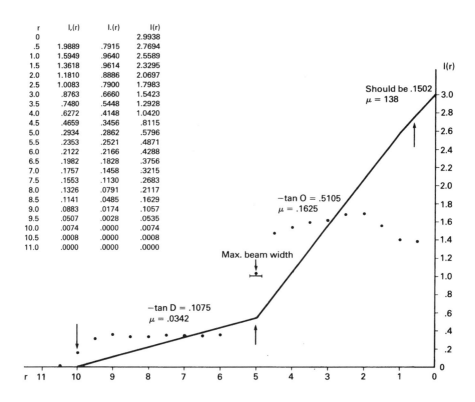

Figure B.2   X-ray reconstruction of a circularly symmetrical object in 1957 [1].

computer. Since he had not had experience of dealing with data in which the principal source of error was statistical, this seemed a good chance for him to learn that too by helping take the data. Final data were taken over two days in the summer of 1963, the calculations were done, and the results are shown in Figures B.4 (a) and (b). Figure B.4 (a) shows the phantom, and this was quite a hybrid. The outer ring of aluminium represents the skull, the Lucite inside it represents soft tissue, and the two aluminium discs represent tumours. The ratio of the absorption coefficients of aluminium and Lucite is about 3, which is very much greater than the ratio of the absorption coefficients of abnormal and normal tissue. The ratio of 3 was chosen in an attempt to attract the attention of those doing what would now be called emission scanning using positron emitting isotopes, a subject then in its infancy. This was about the ratio between the concentration of radioisotope in abnormal tissue and muscle on the one hand and in normal tissue on the other. The mathematics of emission scanning is of course the same as transmission scanning after an obvious correction for absorption. Figure B.4 (b) shows the results as a graph of attenuation coefficient plotted as a function of distance along the line OA. Similar graphs were made along other lines. The full curves are the true values, the dots are the calculated values, and the agreement is quite good. The results could have been presented on a grey scale on a scope, but the graphs were thought to be better for publication.

Figure B.3 Computed tomography scanner used in 1963; its cost was approximately $100.

Publication took place in 1963 and 1964 [1, 2]. There was virtually no response. The most interesting request for a reprint came from the Swiss Centre for Avalanche Research. The method would work for deposits of snow on mountains if one could get either the detector or the source into the mountain under the snow!

My normal teaching and research kept me busy enough so I thought very little about the subject until the early 1970s. Only then did I learn of Radon's [3] work on the line integral problem. About the same time I learnt that this problem had come up in statistics in 1936 at the Department of Mathematical Statistics here in Stockholm, where it was solved by Cramer and Wold [4]. I also learnt of Bracewell's work [5] in radioastronomy which

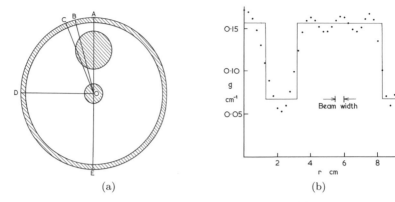

(a)   (b)

Figure B.4   Results obtained with the 1963 CT scanner. (a) Aluminium and lucite
model of a head with two tumours. (b) Solid line: actual values of the attenuation
coefficient along the line OA. Dots: experimental results.

produced the same form of solution as Cramer and Wold had found, of De
Rosier and Klug's [6] (1968) work in electron microscopy, and of the work
of Rowley [7] (1969) and Berry and Gibbs [8] (1970) in optics. Last, but
by no means least, I first heard of Hounsfield and the EMI scanner, about
which you will be hearing shortly. Since 1972 I have been interested in a
number or problems related to the CT scanning problem or stemming from
it, and I would like to tell you about some of them.

As I mentioned earlier, the solution to Radon's problem which I found
was in terms of integrals which extend from the radius at which the absorption coefficient is being found to the outer edge of the sample. What this
means is that if one wishes to find the absorption coefficient in an annulus
one needs only data from lines which do not intersect the hole in the annulus. This is known as the Hole Theorem and it is an exact mathematical
theorem which can be proved very simply. However, as I have mentioned,
when one applies my integrals to real data containing noise, the noise from
the outer observations is amplified badly as one goes deeper and deeper
into the sample, so the integrals as they stand are not useful in practice.
This process of working from the outside in is known as 'peeling the onion',
and several algorithms for peeling the onion have been tried. All show the
same feature, namely that noise from the outer data propagates badly as
one goes deeper and deeper into the sample. However, to the best of my
knowledge, no one has produced a closed form solution to Radon's problem which includes the Hole Theorem and from which it can be proved
that noise from the outside necessarily propagates badly into the interior.

(Doyle and I published a closed form solution to Radon's problem which turned out to be wrong. A retraction will be published soon [9].)

A solution containing the hole theorem for which noise did not propagate into the interior would be extremely valuable for CT-scanning. First it would reduce the dose administered to a patient by not passing X-rays through a region of no interest, that is, the hole. Second it could be used to avoid regions in which sharp changes in absorption coefficient occur. These sharp changes necessarily produce overshoots which cloud the final image. In an extreme case, consider a patient who by accident or by some medical procedure has in his body a piece of metal. This distorts a CT scan of a plane containing the metal beyond recognition. The hole theorem could be used to avoid the piece of metal. Less extremely, an interface between bone and soft tissue introduces overshoots which distort the image of the soft tissue. The hole theorem could be used to avoid the bone and the distortion.

Here then is a question which needs an answer. An unfavourable answer would leave us no worse off than we now are. A favourable answer would result in an improvement of imaging.

In my 1963 paper I had mentioned, but not very favourably, that protons could be used for scanning instead of X-rays. In 1968 my friend and colleague Andreas Koehler started producing some beautiful radiograms [10] using protons, and thinking about his work removed some of the doubts I had previously had about using protons for tomography.

The distinction between protons and X-rays is illustrated in Figure B.5. The number-distance curve for X-rays is the familiar exponential curve. For protons, there is a slow decrease caused by nuclear absorption followed by a very sharp fall-off. Most of the protons penetrate a distance into the material, then all that are left stop in a short distance at what is called the range of the protons. It is this sharp fall-off which makes protons very sensitive to small variations in thickness or density along their paths. A comparison of the two curves at the range is a rough measure of this sensitivity to thickness or density changes. Others were thinking about the same thing and Crowe *et al.* [11] at Berkeley produced a tomogram using alpha-particles in 1975, using the not inconsiderable arsenal of equipment at a high-energy physics laboratory. Koehler and I [12] made a tomogram of a simple circularly symmetrical sample to show two things: (i) that very simple equipment could be used, and (ii) that we could easily detect $\frac{1}{2}$ percent density differences with that equipment. As it happened, we were able to detect density differences of about 0.1 percent or less caused by machining stresses in our Lucite sample and diffusion of water into and out of the Lucite [13]. More recently this work has been continued by Hanson and Steward [14] and their co-workers at Los Alamos, and I am able to show you some of their results. They have recently made tomograms of human organs. Figure B.6 (a) is an X-ray scan of a normal brain made with a A2020 scanner at a dose of about 9 rads. Steward [15] has observed and made density measurements which show that water and electrolyte are lost by white tissue within four hours after death so density differences between grey and white matter disappear. Figure B.6 (b) is a proton scan of the same sample made at a dose of 0.6 rad, and it is roughly as good as the X-ray scan. Figures B.7 (a) and (b) show, respectively, an X-ray scan at a dose of 9 rads and a proton scan at 0.6 rad of a heart with a myocardial infarction fixed in Formalin. According to Steward, fixing in Formalin does not alter the relative stopping powers of the tissues of the heart. Again the pictures are of about the same quality. The time for taking the data was about 1 hour, but the Los Alamos Meson Physics Facility (LAMPF) accelerator is pulsed and is on for only 6 percent of the time. A continuous beam would have produced the same data in about four minutes – roughly the same time as the original EMI scanner. I would like to acknowledge my indebtedness to Hanson and Steward *et al.* for making their results available to me.

Why use protons? First there is the question of dose. The dose in Hanson and Steward's work was about 0.6 rad as distinct from the 9 rads required for the X-ray scans. This is in agreement with theoretical calculations [16] that the dose required for proton scans is five to ten times less than for X-ray scans for the same amount of information. Second, the mechanism by which protons are brought to rest is different from the mechanisms by which X-rays are absorbed or scattered, so one ought to see different things by using the different radiations, particularly the distribution of hydrogen. Koehler and I have indeed found a reversal of contrast between proton and X-ray scans on two occasions, and this is expected from theory.

You know how some people fuss about the high cost of CT scanners, so you can imagine what they would say if one suggested that the X-ray tube in the scanner should be replaced by a much more expensive cyclotron! Of course they would be right, but only if proton scanning is viewed in the narrow context of diagnostics. Instead of looking at the number-distance curve which I showed when comparing X-rays and protons, let us look at the ionization-distance curve which is the thing that matters for therapy. This

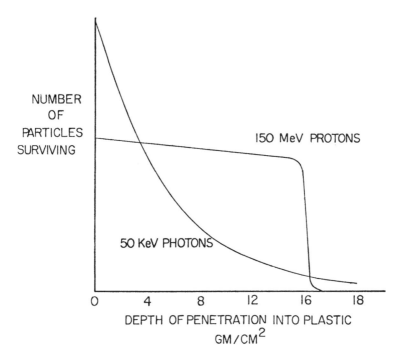

Figure B.5   Number-distance curves for photons (X-rays) and protons penetrating a plastic.

is shown in Figure B.8. You will see that the curve is peaked (the Bragg peak) near the end of the range and then falls off very sharply, so that there is virtually no ionization beyond the end of the range. Both of these facts are of importance in therapy as was pointed out by Wilson [17] in 1946, and as has been shown by treatments of a number of different conditions at Uppsala, Harvard and Berkeley. In fact, protons are far superior to X-rays for the treatment of some conditions. So one can envisage a large metropolitan area as having a 250 MeV proton accelerator with a number of different ports, say ten, at which patients could be treated simultaneously [18]. If one of these ports were devoted to proton tomography, the marginal cost of such tomography would not be great. In exploring these possibilities it is essential that diagnostic radiologists and therapeutic radiologists work together.

Most recently I have been interested in some mathematical topics which I would like to discuss without mathematical details. I have presented

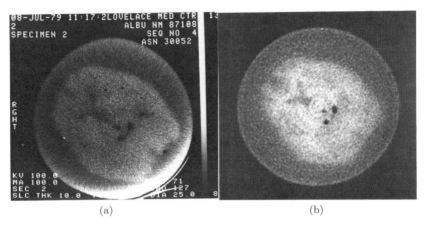

Figure B.6   Scans of a normal brain with (a) X-rays; and (b) protons.

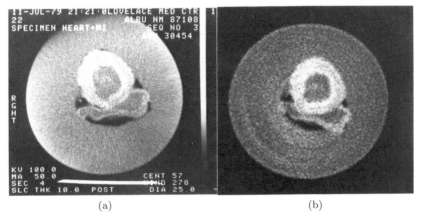

Figure B.7   Scans of a heart with myocardial infarction, fixed in Formalin, with (a) X-rays; and (b) protons.

Radon's problem in the form in which it is usually given in CT scanning – namely, How does one recover a function in a plane given its line integrals over all lines in the plane? This can be generalized to three-dimensions as can be seen in Figure B.9. This shows a sphere in which a function is defined, and a plane intersecting that sphere. Suppose that the given information is the integrals of the function over all planes intersecting the sphere, then the question is "Can the function be recovered from these integrals?" Radon provided an affirmative answer and a formula for finding the function. It turns out that this formula is simpler than in the case of

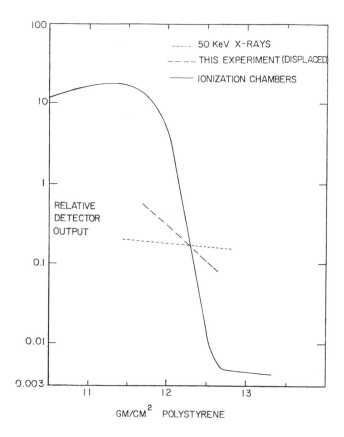

Figure B.8   Ionization as a function of distance for 150-MeV protons penetrating polystyrene.

two-dimensions, and it has applications in imaging using Nuclear Magnetic Resonance (NMR) rather than X-rays. The generalization to n dimensions follows in a straightforward way for Euclidean space and was given by Radon. The few mathematicians who knew of Radon's work before the 1960s applied his results to the theory of partial differential equations. They also generalized his results to integration over circles or spheres of constant radius and to certain ellipsoids. Generalizations to spaces other than Euclidean were made, for example, by Helgason [19] and Gelfand *et al.* [20] in the 1960s.

In my 1964 paper I gave a solution to the problem of integrating over circles of variable radius which pass through the origin as shown in Figure

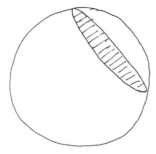

Figure B.9   The hatched region represents a plane intersecting a sphere.

B.10. Recently, Quinto and I [21] have given much more detailed results on this problem generalized to n-dimensional Euclidean space, and we have applied these results to obtain some theorems about the solutions of Darboux's partial differential equation. There is an intimate connection between our results and Radon's results, and we are presently attempting to find more general results which relate the solution of Radon's problem for a family of surfaces to the solution of Radon's problem for another family of surfaces related to the first in a particular way.

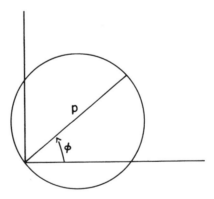

Figure B.10   The polar coordinates (p, ϕ) specify a circle passing through the origin.

What is the use of these results? The answer is that I do not know. They will almost certainly produce some theorems in the theory of partial differential equations, and some of them may find application in imaging with NMR or ultrasound, but that is by no means certain. It is also beside the point. Quinto and I are studying these topics because they are interesting in their own right as mathematical problems, and that is what science is all about.

Of the many people who have influenced me beneficially, I shall name only a few. The late Professor R. W. James F.R.S. taught me a great deal, not only about physics, when I was a student and a young lecturer in his department in Cape Town. Andreas Koehler, director of the Harvard Cyclotron Laboratory, has provided me with friendship, moral support, and intellectual stimulation for over 20 years. On the domestic side are my parents, now deceased, and my immediate family. My wife Barbara and our three children have not only put up with me, they have done so in a loving and supportive way for many years. To these people and others unnamed I shall be grateful to the end of my life.

References

1. Cormack, A. M., J. Applied Physics 34, 2722 (1963)
2. Cormack, A. M. J. Applied Physics 35, 2908 (1964)
3. Radon, J. H., Ber. Sachs. Akad. Wiss. 69, 262 (1917)
4. Cramer, H., and Wold, H., J. London Math. Soc. 11, 290 (1936)
5. Bracewell, R. N., Austr. J. Phys., 9, 198 (1956)
6. De Rosier, D. J., and Klug, A., Nature 217, 30 (1968)
7. Rowley, P. D., J. Opt. Soc. Amer. 59, 1496 (1969)
8. Berry, M. V., and Gibbs, D. F., Proc. Roy. Soc. A314, 143 (1970)
9. Cormack, A. M., and Doyle, B. J., Phys. Med. Biol. 22, 994 (1977); Cormack, A. M., Phys. Med. Biol., to be published
10. Koehler, A. M., Science 160, 303 (1968)
11. Crowe, K. M., Budinger, T. F., Cahoon, J.L., Elischer, V. P., Huesman, R. H., and Kanstein, L. L., Lawrence Berkeley Report LBL-3812 (1975)
12. Cormack, A. M., and Koehler, A. M., Phys. Med. Biol. 21, 560 (1976)
13. Cormack, A. M., Koehler, A. M., Brooks, R. A., Di Chiro, G., Int. Symp. on C. A. T., National Institutes of Health, (1976)
14. Hanson, K., Steward, R.V., Bradbury, J. N., Koeppe, R. A., Macek, R. J., Machen, D. R., Morgado. R. E., Pacciotti, M. A., Private Communication
15. Steward, V.W., Private Communication
16. Koehler, A. M., Goitien, M., and Steward, V. W., Private Communication, and Huesman, R. H., Rosenfeld, A. H., and Solmitz, F. T, Lawrence Berkeley Report LBL-3040 (1976)

17. Wilson, R. R., Radiology 47, 487 (1946)
18. Cormack, A.M., Physics Bulletin 28, 543 (1977)
19. Helgason, S., Acta. Math. 113, 153 (1965)
20. Gelfand, I. M., Graev, M. I., and Shapiro, Z. Y., Functional Anal. Appl. 3. 24 (1969)
21. Cormack, A. M., Quinto, E. T., to appear in Trans. Am. Math. Soc., in press

Appendix C

Presentation of Nobel Prize

On Monday 10 December 1979 Allan Cormack was awarded the Nobel Prize for Physiology or Medicine, an honour that he shared with Godfrey Hounsfield. This is the presentation speech by Professor Torgny Greitz of the Karolinska Medico-Chirurgical Institute. It is followed by the Musical Programme, details of the Solemn Festival and the Menu. All are reproduced here with the permission of the Nobel Foundation.

Your Majesties, Your Royal Highnesses, Ladies and Gentlemen,

Neither of this year's laureates in physiology or medicine is a medical doctor. Nevertheless, they have achieved a revolution in the field of medicine. It is sometimes said that this new X-ray method that they have developed – computerized tomography – has ushered medicine into the space age. Well, sometimes, art can adumbrate reality. In his epic poem about the space ship 'Aniara', Nobel Prizewinner in Literature Harry Martinson tells how, one day, the mimarobe, the computer guardian, "... by means of Mima's formula cycles, phase by phase ... saw into the transtomies ..." and was able to "... see through everything as though it were glass ..."

There, in a single stanza, the poet has captured the essential characteristics and elements of computerized tomography. In addition to X-ray tubes and radiation detectors, the method requires a Mima – that is, a powerful computer; it also calls for a mathematical method, perhaps based on the formula cycles of Fourier transforms, phase by phase; and it produces almost unbelievably clear images of transtomies – that is, cross-sectional views through the human body.

The analogy with the epic poem can be taken even further. The mimarobe remarks: "At this discovery, I went nearly mad ..." Few medical achievements have received such immediate acceptance and met with such unreserved enthusiasm as computerized tomography. It literally swept the world. But the enormous procurement costs involved caused some observers to wonder about the mental health of the health-services sector. Indeed, in the United States, a moratorium on computerized tomography was suggested.

What, then, lay behind this spectacular success? Well, to understand something of the background, let us go back to 1895, the year Röntgen discovered X-rays. The very first X-ray photograph Röntgen ever took – of his wife's hand – indicated, at one and the same time, both the potential and the limitations of conventional X-ray technique. The bones of the hand can be seen, but the complex anatomy of soft tissues – of muscles, tendons, blood vessels, and nerves – does not register.

This inability to distinguish between density differences in the various soft tissues is one of the fundamental limitations of conventional X-ray technique. It means that, in an ordinary X-ray picture, essentially all we can discern are bones and gas-filled spaces. It is the air in the lungs, for example, that enables us to study the lungs' structure and the shape of the heart.

Conventional X-ray technique has two additional shortcomings that are eliminated by computerized tomography.

One such shortcoming is that structures in three-dimensional space overlap in a conventional, two-dimensional, X-ray photograph. What we see is a shadow play – a play, alas, with far too many actors on the stage. It becomes difficult to discern the villain.

The other limitation is that X-ray film cannot indicate any absolute values for variations in tissue density.

Allan Cormack became aware of this latter drawback when, as a young physicist, he was asked to calculate radiation dosages in cancer therapy at Groote Schuur Hospital in Cape Town. He found that the methods then being used – methods based on conventional X-ray examination techniques – were extremely imprecise.

Cormack realized that the problem of obtaining precise values for the tissue-density distribution within the body was a mathematical one. He found a solution and was able, in model experiments, to reconstruct an accurate cross-section of an irregularly shaped object. This was reported in two articles, in 1963 and 1964. Cormack's cross-section reconstructions were the first computerized tomograms ever made – although his 'computer' was a simple desktop calculator.

Cormack realized that his method could be used to produce precise X-ray images and so-called positron-camera pictures of cross-sectional 'slices' of the body. However, no apparatus for practical diagnostic application of these procedures was constructed.

One probable reason for Cormack's difficulty in arousing interest in his experiments was that the computers of the time were incapable of executing – within a reasonable amount of time – the enormous calculations the procedure required.

It was Godfrey Hounsfield who brought Cormack's predictions to fruition. Hounsfield is indisputably the central figure in computerized tomography. Wholly independently of Cormack, he developed a method of his own for computerized tomography and constructed the first clinically usable computerized tomograph – the EMI scanner, which was intended for examinations of the head.

Publication of the first clinical results in the spring of 1972 flabbergasted the world. Up to that time, ordinary X-ray examinations of the head had shown the skull bones, but the brain had remained a gray, undifferentiated fog. Now, suddenly, the fog had cleared. Now, one could see clear images of cross-sections of the brain, with the brain's gray and white matter and its liquid-filled cavities. Pathological processes that previously could only be indicated by means of unpleasant – indeed, downright painful and not altogether risk-free examinations could now be rendered visible, simply and painlessly – and as clearly defined as in a section from an anatomical specimen.

Today, computerized tomography is an established method for the examination of all the organ systems of the body. The method's greatest significance, though, is in the diagnosis of neurological disorders. Since nearly one out of three persons suffers from some disease or disorder of the central nervous system – usually of the brain – during his lifetime, computerized tomography means increased certainty in diagnosis and greater precision in treatment for literally millions of patients.

Cormack and Hounsfield have ushered in a new era in diagnostics. Now they, as well as others inspired by their pioneering contributions, are at work developing yet newer methods for the production of images of cross-sections in the body. In those images, we will be able to discern not only structure, but also function; physiology, or biochemistry. In this, new voyages of discovery are being prepared: voyages into man's own interior, into inner space.

Allan Cormack and Godfrey Hounsfield! Few laureates in physiology or medicine have, at the time of receiving their prizes, to the degree that you have, satisfied the provision in Alfred Nobel's will that stipulates that the prizewinner "shall have conferred the greatest benefit on mankind". Your ingenious new thinking has not only had a tremendous impact on everyday medicine; it has also provided entirely new avenues for medical research. It is my task and my pleasure to convey to you the heartiest congratulations of the Karolinska Institute and to ask you now to receive your insignia from His Majesty, the King.

MUSICAL PROGRAMME

1. FESTIVAL MUSIC AND MELODY	*Edvin Kallstenius*
2. WALTZ FROM "DER ROSENKAVALIER"......	*Richard Strauss*
3. POPULAR BALLAD.......................	*Adolf Wiklund*
4. DANCE OF ZORBA	*Mikis Theodorakis*
5. GREENSLEEVES	*R. Vaughan Williams*
6. HUMORESQUE...........................	*Tor Aulin*
7. OKLAHOMA.............................	*Richard Rodgers*
8. FRAGANCIA.............................	*Evert Taube*
9. BELLMAN-SOIREE	*Karl Nilheim*

Conductor: Karl Nilheim

SONGS FROM "AXION ESTI"
Text by Odysseus Elytis   Music by Mikis Theodorakis
Performed by the choir "Collegium Musicum" from Halmstad
Conductor: Sam Claeson

The students pay homage to the Laureates with participation
of the Stockholm Students' Choral Society
(Stockholms Studentsångarförbund)
Conductor: Lars Blohm

———

The flowers are graciously offered
by the Azienda di Soggiorno e Turismo di Sanremo

SOLEMN FESTIVAL
OF THE NOBEL FOUNDATION

Monday December 10th, 1979, at 4.30 p.m.
in the Grand Auditorium of the Concert Hall

PROGRAMME

"Trumpet Voluntary" .... *Jeremiah Clarke*

The Laureates take their seats on the platform

Speech by *Professor Sune Bergström*, Chairman of the Board of the Nobel Foundation

Ouverture to "Candide" .... *Leonard Bernstein*

Presentation of the Nobel Prize for Physics 1979 to Sheldon L. Glashow, Abdus Salam and Steven Weinberg, after a speech by *Professor Bengt Nagel*

Presentation of the Nobel Prize for Chemistry 1979 to Herbert C. Brown and Georg Wittig, after a speech by *Professor Bengt Lindberg*

Allegretto scherzando from Symphony No. 8 .... L. van Beethoven

Presentation of the Nobel Prize for Physiology or Medicine 1979 to Allan M. Cormack and Godfrey N. Hounsfield, after a speech by *Professor Torgny Greitz*

Pomp and Circumstance No. 4 .... Edward Elgar

Presentation of the Nobel Prize for Literature 1979 to Odysseus Elytis, after a speech by *D.Ph. Karl Ragnar Gierow*, Member of the Swedish Academy

Greek dance No. 5 "Kleftikos" .... Nikos Skalkottas

Presentation of the Bank of Sweden Prize in Economic Sciences in Memory of Alfred Nobel 1979 to Sir Arthur Lewis and Theodore W. Schultz, after a speech by *Professor Erik Lundberg*

The Swedish National Anthem: "Du gamla, du fria"

While the guests are leaving the auditorium, "Festival Music" by *Hugo Alfvén* is played.

Music performed by
the Stockholm Philharmonic Orchestra
Conductor: *Sixten Ehrling*

———

The decoration flowers are graciously offered by
the Azienda di Soggiorno e Turismo di Sanremo

MENU

SUPREMES DE TURBOT FROID AUX OEUFS D'ABLETTE

SAUCE HOLLANDAISE

SELLE DE VEAU ROTIE

SAUCE AUX MORILLES

POMMES SAUTEES

SALADE DE HARICOTS VERTS

PARFAIT GLACE NOBEL

PETITS FOURS

———

VINS

G.H. MUMM, CORDON ROUGE, BRUT

CHATEAU TRINITE-VALROSE 1970

EAU MINERALE DE RAMLÖSA

———

CAFÉ

BARON OTARD, FINE CHAMPAGNE, V.S.O.P.

LIQUEUR MANDARINE NAPOLEON

———

BUFFET

BOLS SILVER TOP DRY GIN

LONG JOHN WHISKY

Appendix D

Man and Science in the 21st Century

In January 1984 Allan Cormack received a letter from Daisuke Yamauchi, President of *The Mainichi Newspapers*, inviting him to contribute an essay to "give some hints to those who plan to apply for the thesis contest" on considering the 21st century. The essay was readby over 4 million Japanese but has never before been published in English. It is reproduced here with the permission of the publisher.

In teaching elementary physics it is important to stress the effects which the finite speed of light has on the way we look at life. I like to tell my students that this not only forbids us from predicting the future (since no light signals can be received by us from events in the future), but it also prevents us from saying what is happening now at distant parts of the universe. For example, if a star is five light-years from where you were at the instant you read this, you would be unable to know anything about it until at least five years from now. Having established that according to present day physics there is no way of predicting the future, let me now proceed to guess what the relation between Man and Science will be in the 21st century. My guessing will be based on an extrapolation of the increasingly rapid development of science over the last few hundred years and to understand this it is useful to take a brief look at the development of human thought.

As we know from early myths and early written records, man has long tried to understand his place in the universe. Initially this seemed to invoke many gods and spirits with highly capricious natures to 'explain' the highly capricious nature of life on earth. As time went on, man's thoughts became

more systematized, and for at least 2500 years we have been able to talk
about man having a philosophy or philosophies of life. By philosophy I mean
here a coherent attempt to understand life in all its aspects. On the one
hand philosophy should account for the brute facts of life, for example the
existence of the Earth, the Sun, the Moon and the stars or, at a seemingly
more commonplace level, the fact that a stone thrown into the air returns
to the ground, or the fact that a particular type of clay fired in a particular
way will produce a pot. On the other hand, there are man's perceptions
of himself, of other men and of their interaction with nature and with
themselves. Here we are involved in questions of aesthetics, politics, ethics,
and, of course, religion. In the western world one thinks of many people
starting with Plato and Aristotle who attempted such grand syntheses.

Until quite recently all philosophies were severely limited by the absence of what we now call science. Apart from a brief brilliant flourishing
among the Ionian philosophers about 600 BC (which put the Earth, Sun
and Moon in some kind of sensible order, and which was quickly forgotten)
we had to wait until Copernicus, Galileo and Newton before we could even
understand why a stone which was thrown into the air would return to the
ground. These men and others like them ushered in the subject of Natural
Philosophy (*i.e.* the study of natural phenomena) as distinct from Moral
Philosophy, which covered the rest of philosophy. Starting only about 300
years ago Natural Philosophy grew with ever increasing speed and soon
reached the point where no one person could understand it all, and it was
fragmented into Physics, Mathematics, Chemistry, Botany, Zoology, Geology, Psychology and so on. On the whole, these studies did not conflict
with Moral Philosophy very much (though there were notable exceptions,
for example, the trial of Galileo) and they could be brought into harmony
with revealed religion by an enlightened clergy. There was one great exception, the work of Darwin. Until Darwin, each tribe, people, religious
group, had its own creation myth which had a deity or deities which delivered over to them (and them only!) the land which they occupied. Darwin
had the temerity to suggest not only that man had evolved from 'lower'
species, but also to provide a mechanism (natural selection) which worked
through competitive behaviour to secure the survival of the fittest. This
had and still has a profound influence not just on religion, but on the whole
of Moral Philosophy, and it is important to remember that Darwin's *Origin
of Species* was published only 125 years ago. With Darwin the fragmentation of Natural Philosophy seemed complete and hence the fragmentation
of Philosophy itself was complete. For a while, it seemed that it was no

longer possible to build a great system of philosophy, say like Aristotle's, and I have heard philosophers lament this fact. What I am going to suggest is that we are on our way back to a grand synthesis, that this synthesis will be well under way in the 21st century and that this synthesis will be achieved by subsuming almost all of Philosophy in Science.

To see why I believe that this may happen, let us look at a number of things which have happened in my lifetime, some in the physical sciences and some in biology and anthropology.

While most philosophies have included some cosmology it was only sixty years ago that it was realized that, first, our Sun was only one star among many in a galaxy, and second and more important, that many little bits of nebulosity which astronomers saw in the heavens were in fact other galaxies, remote from our own and from each other. We were just starting to learn that these other galaxies are receding from us. Today we know that some of these galaxies are enormously remote from us, and that light which we are receiving from them left billions of years ago, so that we are seeing matter as it was in the early stages of the universe. This analysis of the motion of these galaxies has led us to believe that the observable universe started with the Big Bang about ten billion years ago. At the moment, we are unable to tell whether it will start contracting again some time, possibilities which were suggested over sixty years ago.

As has been mentioned, gravity was first explained by Newton, and electricity and magnetism were fused into one subject in the 19th century, but we had to wait until the 1930s to see clearly that there were two other fundamental and quite different types of interactions in nature: the strong and the weak interactions. Since the 1930s attempts to understand these, and to understand them in a way which is consistent with the laws of quantum mechanics and relativity, made slow progress until about fifteen years ago, when the grand unified theories of Glashow, Salam and Weinberg achieved the fusion of the electromagnetic and the weak interactions and showed promise of including the other two.

This by itself was a notable achievement, but alongside it developments took place in astrophysics, nuclear physics and geophysics. Together these have provided an amazing synthesis of the birth and development of the universe, of the birth and death of galaxies and stars, of the formation of the elements, and even why the earth is as it is. I am not suggesting that the synthesis is complete or that there are not many fascinating questions to be answered, or that present interpretations of the facts may not change, but I am suggesting that this synthesis in a single quantitative framework would boggle the minds of even the greatest philosophers of the past.

Biology has not been standing still. Man's ancestry has been credibly traced back through 2 million years. Many chemicals which were at one time thought to be obtainable only in nature have been synthesized. With the understanding of the structure and function of DNA and related substances we are now in a position to understand the propagation of species including our own. While there is still some argument about the mechanisms and rates of speciation, Darwin stands vindicated, and evolution is receiving assistance from genetics and biochemistry. So, for example, biochemistry has shown that among the higher apes the chimpanzees are our closest relatives.

The pioneering work of Konrad Lorenz, Karl von Frisch and Nikolaas Tinbergen has provided us, over the past 60 years, with an entirely new biological science: ethology, the study of animal behaviour. By studying the behaviour of such diverse creatures as ants, bees, geese, seagulls and, more recently, chimpanzees and gorillas and other mammals we are now able to ask serious questions about such things as altruism. Is a bee which sacrifices itself in defence of the hive behaving in a way which humans would call 'altruistic', or are there other explanations, such as the preservation of the gene pool of the hive? This subject is in a very early stage so that there are few questions which have firm answers and, as in the early stages of any subject, it is not clear which are the right questions, but we should rejoice that we are asking questions and expecting answers. I firmly believe that these studies, together with a lot of physiological and biological studies, will ultimately lead to a better understanding of human nature than we now have. This is of course very disturbing to those who have inherited or divined strong beliefs about the nature of man, and the hostile reception given to E.O. Wilson's attempt at developing a new subject called Sociobiology is reminiscent of the attacks on Galileo and Darwin. I am not saying that Wilson is the Galileo or Darwin of this subject, but it is my belief that if he is not, some new Galileo or Darwin will arise and establish a new science along these lines. In this new subject, whatever it is, I believe that ethics, politics and even aesthetics will be subsumed along with the cognitive sciences in physiology, biochemistry and genetics, and thus, ultimately, in physics.

Assuming that this grand synthesis takes place, how will it affect us? How will we regard art, literature, and music, for example? It is hard even to guess. On the other hand, we may have a deeper and more profound appreciation of the great classics, including those still to be created. On the other, art, literature and music as we now know them may be swept away and replaced by something entirely new, and in the appreciation of this new thing we may experience more profoundly than ever before a sense of oneness with the universe which is the ultimate achievement in some mystical religions.

Two profound questions will remain. One is the question of ultimate cause. For example, if present day cosmology remains unaltered, the question is "What caused or preceded the Big Bang?" Physics has no answer to this today, and it is difficult to see where an answer may be found in physics. The other question is "How is it that we, among all species, are able to separate ourselves sufficiently from our surroundings to be able to think about them?" Is this something which uniquely separates us from even our nearest cousins, the chimpanzees, or is it a natural part of evolution? If it is a part of evolution, then when and how did it happen to us in the last two million years? Are chimpanzees dimly aware of the need to explain the world around them and would they develop habits of thought like ours if they were left to evolve along their own lines? These are hard questions indeed.

From its beginnings science has always had a dark side, namely, its application to the destruction of other human beings. Archimedes is supposed to have used scientific methods of warfare in defence of Syracuse against the Romans. No sooner were Galileo's early studies in mechanics published than they were applied to calculate trajectories of cannon-balls, and today we live in constant dread of nuclear war. Unless this dark side of science is brought under control, we may never even see the beginning of the 21st century. The world and all that is in it may go down to a flaming destruction, as in the old Germanic myths, instead of being something like the paradises envisioned in various religions.

Appendix E

A Teenager's Odyssey

In December 1991 Allan Cormack was visited by Hiroshi Goto, editor of *The Weekly Shonen Jump*, a magazine for teenagers. He invited Allan to contribute a message for the young readers "to renew their hopes for the future". The short essay, followed by the story of Allan's life, was read by many millions of Japanese but has never before been published in English. It is reproduced here with the permission of the publisher.

This is a story about the physicist who succeeded in coalescing X-rays and computers. As a 14-year-old boy, Cormack was brought up and surrounded by the beautiful scenery of South Africa.

For medical treatment, accurate diagnosis is essential. The birth of the CAT scanner has created a revolution, a huge breakthrough in methods of diagnosis. The physicist who predicted and theorized this epoch-making high technology reflects on his childhood and the message he has for you, the young people of today, is ...

When you watch television or when you play computer games, do you ever wonder how the rapidly changing pictures are formed? A fine beam of electrical 'bullets', called electrons, scans the screen tens of thousands of times a second. When the electrons strike the screen they cause the atoms of the screen to emit 'bullets' of light called photons. When these photons enter your eye they send messages to your brain which cause you to see the pictures. You probably don't know much about electrons and photons and atoms, and perhaps you are even a little afraid of the words. If you are afraid you shouldn't be. Only a hundred years ago not a single person in the whole world even knew the words 'electron' and 'photon', because electrons and photons hadn't yet been discovered, and only a few people in the world took atoms seriously. Now we know enough about these things (and a lot of other particles) so that we can build television sets and computers and so on, but there is still a lot we don't know. Science tries to explain what we don't understand, and tries to fit that explanation into our other explanations of everything else in the Universe: stars, galaxies, and even life itself.

Thinking about becoming a scientist can be quite scary. An awful lot has been discovered in the past, and new discoveries are being made every day, so how can you possibly learn all that stuff before you reach the point when you can make scientific discoveries? That is where science is beautiful: you don't have to learn all that stuff! Each new theory to explain new facts must explain all the old theories and the old facts too! This is why people, especially young people, can soon reach the point when they can make valuable discoveries if they are prepared to make the effort.

You have to be prepared to make the effort, that is, to work hard. Even the greatest scientists have to work quite hard to make their discoveries, so it is reasonable to ask what rewards you might expect for your hard work if you become a scientist. I must say right away that you cannot expect to make a lot of money. Only a very few scientists have made a lot of money out of their discoveries, and in all cases they have been people who have found practical, profitable applications for their work. So why do people become scientists? They must be very curious, so that they simply cannot rest until they have found an explanation for some problem which is bugging them. Finding an explanation is a wonderful experience, like creating a work of art, and it is terribly addictive! Once you have solved one problem you cannot wait to get on to the next one! I would like you all to think about aiming to become a scientist. When I look back at the 68 years that constitute my life, I cannot think of any other job more satisfying than becoming a scientist.

* * * * * *

I was born in 1924 in Johannesburg, South Africa. My father was a telephone engineer who worked at the post office. Due to the fact that my father had many transfers for his job, my family and I moved to Cape Town when I was eight. When I was six, my father had been infected with pneumonia. Since there were no antibiotics for it at the time, the chances even for adults to live were slim, but luckily he survived. However, ever since the infection, he would get sick very easily and finally when I was 12, he died young due to renal failure. The death of my father was the beginning of some sad bitter days.

During World War II, it was illegal to take any pictures in the city of Cape Town, but I managed to save one picture from that time. Maybe due to the fact that I was the youngest child in my family, there was a period of time, more or less after my father's death, when I hated school. I stopped going to school, so my mother took me out for a vacation! Although we didn't have much money, my mother, who was a school teacher, decided to take me to Scotland, where she was born. My mother never told me to "Go to your room and study" or "I want you to work hard and get good grades". My mother's hometown in Scotland was beautiful and when we came back to South Africa, I naturally started going to school again. "Go ahead and do something that makes you the most happy", is what my mother would always say. I have always kept these words in my life.

In South Africa, there are many snakes that are called the Common Brown Snake. At school I would always grab them, stuff them in my pocket and then attend the morning assemblies. I knew that everybody was gathered for the morning assembly so there was nobody in the classroom. As soon as the assembly was over, I would run back to the classroom before anybody got back, stuff the snakes into my teacher's briefcase and wait for everybody to return. When the teacher came back and opened his briefcase, poof! The teacher would scream "Get the snakes!" and the kids would panic. I have always played some tricks like this, but I never got into trouble for it. Rather, the teachers would always have fun with me. I was very fortunate to have great teachers.

Cape Town is where I spent my early life. In this city, where I also spent my student years, there is a beautiful mountain called Table Mountain, which is named after its shape. Table Mountain is perfect for rock and mountain climbing, since it doesn't get icy nor does it get too much snow and the view when climbing it is phenomenal. The rocks of Table Mountain are undoubtedly the most beautiful in the world, and I still sometimes dream about it today. I also enjoyed skiing as much as I enjoyed climbing. When the temperature would drop down and if I saw frost on the top of the mountains, I would always call a farm house that's located nearby to check if there was enough snow for skiing. If there was enough snow, I would skip college and go skiing with my friends.

In high school, my favorite subject was physics. Going to the library became an everyday habit, and I began reading books on astronomy. I learned more about the beauty of astronomy by reading books by James Jeans, who was a physicist as well as an astronomer, and by the eminent Arthur Stanley Eddington. From that time I really wanted to pursue astronomy and do something with it in my later life. The more I read the more I realised that in order to become an astronomer I would need to perfect my mathematics and physics, which is exactly what I did. It was so interesting, I became addicted. However, I soon realised that to get a job that had to do with astronomy in South Africa, would be almost impossible unless you were a genius. So then, with a scholarship, I entered the faculty of electronic engineering at the University of Cape Town. A professor who served as a technical advisor during World War II came back to the university to teach and explained that the curriculum for the faculty needed to be shifted to more mathematics and physics. This is when I decided to change my major from engineering to physics.

I was right with my choice of changing my major. I liked it so much that I got myself a master's degree in physics. With another scholarship, I studied abroad, at Cambridge University in Great Britain. The reason why I wanted to study at Cambridge was because I wanted to study nuclear physics under Professor Rutherford, a Nobel Prize winner. However, by the time I got there, Professor Rutherford was no longer at the university: he had died a few years before. Although I didn't mention it to anybody, I was depressed.

During my stay at Cambridge University, life was hard but studying was fun. I met many lifetime friends, girlfriends, and even the special girl who would become my wife. I remember once when I was at the train station in London I heard an announcement that this girl wanted me to visit her in America!

Although I wanted to find a job in Great Britain, it didn't go as smoothly as expected, but fortunately when I came back to Cape Town, I was able to find a job as a lecturer in the physics department. During the time when my wife and my children were in Cape Town, I was offered a part-time job in the radiology department at the hospital affiliated with the University of Cape Town. My predecessor got sick of South Africa and left, and I was the only nuclear physicist in the country at the time.

Every weekend, I would commute to the hospital and help out with the radiotherapy treatments. Since I wasn't a medical student, I was not used to looking at pictures of X-rays and I had a hard time. While I was helping out, I was shocked to see the carelessness of radiotherapy treatment that was taking place. The bones are hard, but the internal organs are soft. Even within the internal organs, there is oxygen within respiratory organs, but not within the liver. However, they were all being treated via the same method. I was surprised at how sloppy it was! Even as an outsider I couldn't help but wonder if this wasn't too inaccurate. Treatment of a patient by irradiation was based on the premise that all human bodies are the same. The fact that each body part needed different amounts of radiation was quite obvious to me. How much radiation was needed for the treatment? As a physics researcher, I was able to solve this question by calculating and accurately mapping the amount necessary.

As a university lecturer and helper at the hospital, nearly six years flew by. The following year, I was eligible to take sabbatical leave with full pay. Since my wife was an American from Boston and had studied at Harvard University, she had many acquaintances there. She advised me to write a letter to Professor Ramsey. To my surprise, I received a positive response from him, and this even helped me realize what I wanted to do with my life. I had been saved by my wife's words! Since I was on sabbatical, my plan was to leave my post in Cape Town for about a year or so. I decided to join the research group that Professor Ramsey ran. During the course of the year, the experiments went well, but by the end I realized that I wanted to do more research. Luckily, I was offered a job teaching at Tufts University which is located right next door to Harvard University. After returning to South Africa for a few months, I said farewell to Cape Town and my beloved Table Mountain.

After working at Tufts University for a while, I decided to continue with the mapping and calculations that I had started when I was in Cape Town. Although I only had time for this research in my spare time, I was finally able to map out the cross-sectional image of a body scanned by X-rays from different dimensions using a computer. In 1963 and 1964 I released these findings to the public. At this time, I was also able to explain mathematically the function of an MRI scanner, where nuclear magnetic resonance is used to create images. However, not many people paid particular attention to this accomplishment. I also realised this about scientific accomplishment: if it's too early, it will be ignored; if it's too late, it's too late!

Later on, Godfrey Hounsfield from the United Kingdom offered many

doctors the opportunity to use a CAT scanner, since this machine allowed them to see a visual image of the inside of the skull. With the CAT scanner, the doctors were easily able to detect a tumour in the brain. This was a miracle since the examinations before this invention meant that doctors were just groping about in the dark to see what they could find. "This is spectacular", the doctors would say. This is when things started to change. Everyone was interested. This was the revolution of the CAT scanner!

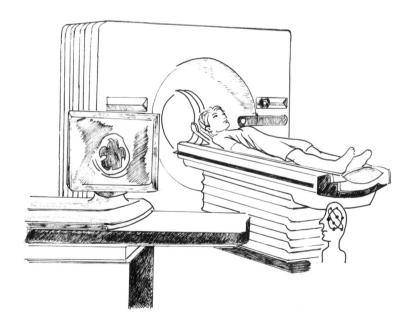

The machine that Hounsfield invented only allowed us to capture pictures of the skull, but later with my mathematical explanation, we were able to come up with a new machine that could take pictures of other body parts as well. The demand for this revolutionary diagnostic device spiked up in hospitals all over the world. The MRI was also invented and contributed greatly to the world of medicine. In order to treat a disease, an accurate diagnosis is crucial. The main goal of my research has been to prove that this was the right way to diagnose. People might be surprised when they hear what I had been thinking about during my experiments. For me, it was pure coincidence that my research project resulted in such success, but some people may say it was just my strong will to keep exploring and investigating.

* * * * * *

Dr Cormack won the 1979 Nobel Prize in Physiology or Medicine for his theories together with Godfrey Hounsfield, who invented the machine which allowed computers to produce three-dimensional graphic images based on computed tomography of X-rays. Dr Cormack's theory was also in accordance with the later discovery of MRI, and it greatly contributed to the world of medical science. He studied abroad in Britain, came back to South Africa to be a teacher at the University of Cape Town, and then became a research student at Harvard University. In 1957 he became a professor at Tufts University as well as a researcher. He has been elected as a research fellow of radiological institutions all over the world. Not only is he a well known physicist, he has also become an expert on radiation therapy treatment. In 1966, he became a naturalised citizen of the United States of America.

Notes

Chapter 1   From John O'Groats to Jo'burg

1. Cormack AM, 'Some family history', unpublished notes, Winchester, MA, page 2, 1996.
2. Glover JR, *The Story of Scotland*, Faber & Faber, London, page 15, 1977.
3. Cormack AM, 'Some family history', unpublished notes, Winchester, MA, page 6, 1996.
4. Glover JR, *The Story of Scotland*, Faber & Faber, London, pages 17-18, 1977.
5. Cormack AM, 'Some family history', unpublished notes, Winchester, MA, page 6, 1996.
6. Glover JR, *The Story of Scotland*, Faber & Faber, London, page 38, 1977.
7. Cormack AM, 'Some family history', unpublished notes, Winchester, MA, page 7, 1996.
8. Glover JR, *The Story of Scotland*, Faber & Faber, London, page 188, 1977.
9. Cormack AM, 'Some family history', unpublished notes, Winchester, MA, page 7, 1996.
10. Cormack AM, 'Some family history', unpublished notes, Winchester, MA, page 8, 1996.
11. Heritage Centre, Wick, Caithness, Scotland, 14 June 2006.
12. Cormack AM, 'Some family history', unpublished notes, Winchester, MA, page 9, 1996.
13. Heritage Centre, Wick, Caithness, Scotland, 14 June 2006.
14. Cormack AM, 'Some family history', unpublished notes, Winchester,

MA, page 10, 1996.

15. http://en.wikipedia.org/wiki/Cormac_mac_Airt

16. 'The death of Cormac mac Airt', *Annals of the Four Masters*, County Donegal, 1636.

17. O Hogain D, *Myth, Legend and Romance. An Encyclopaedia of the Irish Folk Tradition*, Prentice Hall, London, pages 121-126, 1991.

18. http://en.wikipedia.org/wiki/Cormac_mac_Airt

19. Cormack AM, 'Some family history', unpublished notes, Winchester, MA, pages 9–10, 1996.

20. Letter from George Cormack to his grandson George Cormack, 30 December 1903.

21. Interview with Amy Read, 20 July 2004.

22. Thomson LM, 'South Africa', *The World Book Encyclopedia*, World Book, Chicago, Volume 18, page 619, 1988.

23. Coffman EM, 'World War I', *The World Book Encyclopedia*, World Book, Chicago, Volume 21, page 454, 1988.

24. Cameron T, *Jan Smuts: An Illustrated Biography*, Human & Rousseau, Cape Town, 1994.

25. Potgieter DJ, *The Standard Encyclopaedia of Southern Africa*, Nasou, Cape Town, pages 245–246, 1973.

26. Thomson LM, 'South Africa', *The World Book Encyclopedia*, World Book, Chicago, Volume 18, page 620, 1988.

27. Biographical archive for Allan MacLeod Cormack, St John's College, Cambridge, June 2006.

28. Interview with Amy Read, 13 July 2004.

29. Rosenberg V, *The Life of Herman Charles Bosman*, Human & Rousseau, Cape Town, 1991.

30. Type-written caption in Allan Cormack's photo album, 1996.

31. Jabavu DDT, *The Segregation Fallacy*, Lovedale Institution Press, Eastern Cape, 1928.

32. Thomson LM, 'South Africa', *The World Book Encyclopedia*, World Book, Chicago, Volume 18, page 620, 1988.

33. Interview with Amy Read, 9 August 2006.

34. Interview with Amy Read, 3 April 2006.

35. Phillips H, *The University of Cape Town 1918-1948. The Formative Years*, UCT Press, Cape Town, page 425, 1993.

36. Interview with Amy Read, 13 July 2004.

37. Interview with Amy Read, 9 August 2006.

38. Interview with Amy Read, 20 September 2004.

39. Hand-written notes in Allan Cormack's photo album, 1938.
40. Interview with Amy Read, 20 September 2004.
41. Stokesbury JL, 'World War II', *The World Book Encyclopedia*, World Book, Chicago, Volume 21, page 470, 1988.
42. Hand-written notes in Allan Cormack's photo album, 1938.
43. 'The physicist who succeeded in coalescing X-rays and computers' (in Japanese), *Shonen Jump*, 33: 249–255, 1992.
44. Interview with Amy Read, 20 September 2004.
45. Cormack AM, 'Some family history', unpublished notes, Winchester, MA, page 2, 1996.
46. Hand-written notes in Allan Cormack's photo album, 1938.
47. Interview with Amy Read, 20 September 2004.

Chapter 2   On the Slopes of Table Mountain

1. E-mail letter from Allan Cormack to John Wanklyn, 1 March 1998.
2. Cormack AM, Letter published in Rondebosch Old Boys' Newsletter, September 1979.
3. Veitch N, *Rondebosch Boys' High and Preparatory Schools 1897–1997*, Published by the Centenary Committee, Cape Town, page 78, 1996.
4. Cormack AM, 'Some family history', unpublished notes, Winchester, MA, page 2, 1996.
5. Veitch N, *Rondebosch Boys' High and Preparatory Schools 1897–1997*, Published by the Centenary Committee, Cape Town, page 78, 1996.
6. Cormack AM, 'Some family history', unpublished notes, Winchester, MA, page 2, 1996.
7. Veitch N, *Rondebosch Boys' High and Preparatory Schools 1897–1997*, Published by the Centenary Committee, Cape Town, page 79, 1996.
8. Cormack AM, Letter published in Rondebosch Old Boys' Newsletter, September 1979.
9. Veitch N, *Rondebosch Boys' High and Preparatory Schools 1897–1997*, Published by the Centenary Committee, Cape Town, page 69, 1996.
10. Veitch N, *Rondebosch Boys' High and Preparatory Schools 1897–1997*, Published by the Centenary Committee, Cape Town, page 81, 1996.
11. Cormack AM, 'Autobiographical Sketch', unpublished notes, Winchester, MA, page 1, 8 June 1990.
12. Crowther JG, 'Arthur Stanley Eddington 1882–1944', *British Scientists of the Twentieth Century*, Routledge & Kegan Paul, London, pages 140–196, 1952.

13. http://www-groups.dcs.st-and.ac.uk/~history/Mathematics/ Eddington.html

14. Crowther JG, 'James Hopwood Jeans 1877–1946', *British Scientists of the Twentieth Century*, Routledge & Kegan Paul, London, pages 93–139, 1952.

15. http://www-groups.dcs.st-and.ac.uk/~history/Mathematics/ Jeans.html

16. Stokesbury JL, 'World War II', *The World Book Encyclopedia*, World Book, Chicago, Volume 21, page 470, 1988.

17. Thomson LM, 'South Africa', *The World Book Encyclopedia*, World Book, Chicago, Volume 18, page 620, 1988.

18. *RBHS Magazine*, Vol. XXXII, No. 124, page 1, November 1939.

19. *RBHS Magazine*, Vol. XXXIV, No. 127, page 13, June 1941.

20. *RBHS Magazine*, Vol. XXXIII, No. 125, page 1, June 1940.

21. *RBHS Magazine*, Vol. XXX, No. 120, page 19, December 1937.

22. *RBHS Magazine*, Vol. XXIX, No. 118, page 14, September 1936.

23. *RBHS Magazine*, Vol. XXXIV, No. 128, page 14, December 1941.

24. Veitch N, *Rondebosch Boys' High and Preparatory Schools 1897–1997*, Published by the Centenary Committee, Cape Town, page 34, 1996.

25. Veitch N, *Rondebosch Boys' High and Preparatory Schools 1897–1997*, Published by the Centenary Committee, Cape Town, page 57, 1996.

26. *RBHS Magazine*, Vol. XXXIV, No. 127, page 9, June 1941.

27. 'Scholars' Fine Performance', *Cape Argus*, 17 June 1941.

28. Interview with Amy Read, 13 July 2004.

29. *RBHS Magazine*, Vol. XXXIV, No. 127, page 9, June 1941.

30. *RBHS Magazine*, Vol. XXXIV, No. 127, page 5, June 1941.

31. *RBHS Magazine*, Vol. XXXV, No. 129, page 6, June 1942.

32. Cormack AM, 'Autobiographical Sketch', unpublished notes, Winchester, MA, page 2, 8 June 1990.

33. Stokesbury JL, 'World War II', *The World Book Encyclopedia*, World Book, Chicago, Volume 21, page 480, 1988.

34. Allan Cormack's Registration Form, No. 815, University of Cape Town, 27 February 1942.

35. Official Transcript of Courses Taken by Allan Cormack, University of Cape Town, 1942–1945.

36. 'Shirking Our Duty?' *Varsity*, UCT Student Newspaper, page 1, 15 May 1942.

37. 'Principal Speaks', *Varsity*, UCT Student Newspaper, page 3, 29 May 1942.

38. Phillips H, *The University of Cape Town 1918–1948*, UCT Press, Cape Town, page 298, 1993.
39. Phillips H, *The University of Cape Town 1918–1948*, UCT Press, Cape Town, page 300, 1993.
40. Interview with Godfrey Stafford, 15 October 2004.
41. Cormack AM, 'Autobiographical Sketch', unpublished notes, Winchester, MA, page 3, 8 June 1990.
42. Annual Prospectus, Faculties of Arts and Science, University of Cape Town, pages 57–104, 1943.
43. 'Committee of Engineering Society', *Varsity*, UCT Student Newspaper, page 4, 6 May 1943.
44. 'Engineering Society', *Varsity*, UCT Student Newspaper, page 3, 6 May 1943.
45. Official Transcript of Courses Taken by Allan Cormack, University of Cape Town, 1942–1945.
46. Cormack AM, 'Autobiographical Sketch', unpublished notes, Winchester, MA, page 3, 8 June 1990.
47. Cherry RD, 'Professor RW James', *South African Journal of Science*, 93: 94–96, 1997.
48. Phillips H, *The University of Cape Town 1918–1948*, UCT Press, Cape Town, page 344, 1993.
49. Cormack AM, 'Autobiographical Sketch', unpublished notes, Winchester, MA, page 3, 8 June 1990.
50. Interview with Brian Warner, 15 March 2006.
51. Stoy RH, Cormack A, 'On the photometric use of plates taken for the determination of stellar parallax', *Monthly Notices of the Royal Astronomical Society*, 105(4): 225–236, 1945.
52. 'Professor James Sights the Atom', *Varsity*, UCT Student Newsletter, page 3, 6 June 1945.
53. Stokesbury JL, 'World War II', *The World Book Encyclopedia*, World Book, Chicago, Volume 21, page 494, 1988.
54. Versveld M, 'Bomb Condemned', *Varsity*, UCT Student Newsletter, page 3, 27 August 1945.
55. Official Transcript of Courses Taken by Allan Cormack, University of Cape Town, 1942–1945.
56. Cormack AM, 'Autobiographical Sketch', unpublished notes, Winchester, MA, page 3, 8 June 1990.
57. Scholtz DL, 'The magmatic nickeliferous ore deposits of East Griqualand and Pondoland', *Transactions of the Geological Society of South*

Africa, 39: 81–210, 1936.

58. Cormack AM, 'X-ray powder data of the Parkerite series $Ni_3Bi_2S_2$ – $Ni_3Pb_2S_2$', *Transactions of the Geological Society of South Africa*, 50: 17–22, 1947.

59. Phillips H, *The University of Cape Town 1918–1948*, UCT Press, Cape Town, page 347, 1993.

60. Interview with Godfrey Stafford, 15 October 2004.

61. Cherry RD, 'Professor RW James', *South African Journal of Science*, 93: 94–96, 1997.

62. Interview with Amy Read, 20 July 2004.

63. Interview with Walter Powrie, 26 November 2006.

64. Carter D, 'Moving up and on from a rope', *Mountain Ears, Newsletter of the Mountain Club of South Africa*, Cape Town, page 10, June 2004.

65. Biesheuvel H, Cormack A, 'Procrastination Crag', *Journal of the Mountain Club of South Africa*, pages 70–73, 1946.

66. de Lisle M, 'Cormack's Route', Personal Logbook, 21 July 1946.

67. Interview with Michael de Lisle, 10 September 2004.

68. de Lisle M, 'Waaihoek', Personal Logbook, 9–16 July 1945.

69. Norton M, 'To Mariana, On Her Fall', Poem in the Personal Logbook of Michael de Lisle, 13 July 1945.

70. 'Two Cape Town Mountain Climbers Frozen to Death', *Cape Times*, 13 August 1945.

71. de Lisle M, Personal Logbook, 14 August 1945.

72. de Lisle M, 'Cedarberg', Personal Logbook, 13–22 December 1945.

73. Letter from Cecilia (de Lisle) Crofts to Jean Cormack, 26 October 1999.

74. Interview with Michael de Lisle, 10 September 2004.

75. Phillips H, *The University of Cape Town 1918–1948*, UCT Press, Cape Town, page 240, 1993.

76. de Lisle M, 'Varsity Rag', Personal Logbook, 23 March 1946.

77. Phillips H, *The University of Cape Town 1918–1948*, UCT Press, Cape Town, page 240, 1993.

78. Phillips H, *The University of Cape Town 1918–1948*, UCT Press, Cape Town, pages 255–256, 1993.

79. 'Degree for Her Majesty the Queen', *Varsity*, UCT Student Newspaper, page 1, 7 March 1947.

80. Klug A, 'Crystal structure of p-bromochlorobenzene', *Nature*, 160: 570, 1947.

81. Interview with Aaron Klug, 18 October 2004.

82. http://nobelprize.org/nobel_prizes/chemistry/laureates/1982

Chapter 3   Physics and Friends at Cambridge

1. Cormack AM, 'Autobiographical Sketch', unpublished report, Winchester, MA, page 4, 8 June 1990.
2. Raphael F, 'Return of the Cambridge native', *Sunday Times*, 15 January 2006.
3. Jones B, Dixon MV, 'Sir Isaac Newton', *The Macmillan Dictionary of Biography*, Macmillan, London, page 578, 1982.
4. Raphael F, 'Return of the Cambridge native', *Sunday Times*, 15 January 2006.
5. James RW, Register of 20th Century Johnians, St John's College, Cambridge, June 2006.
6. Theron H, 'Shackleton's odyssey recounted in diary', *Monday Paper*, University of Cape Town, 22(33):1, 3 November 2003.
7. Raphael F, 'Return of the Cambridge native', *Sunday Times*, 15 January 2006.
8. Cormack AM, 'Autobiographical Sketch', unpublished report, Winchester, MA, page 4, 8 June 1990.
9. Letter from Allan Cormack to Phillip Tobias, 17 April 1985.
10. Lindsey G, 'A colonial research student at post-war II Cambridge', unpublished report, 26 September 2004.
11. Interview with Godfrey Stafford, 15 October 2004.
12. Cormack AM, 'Autobiographical Sketch', unpublished report, Winchester, MA, page 5, 8 June 1990.
13. http://www2.mrc-lmb.cam.ac.uk/archive/nobel.html
14. Lindsey G, 'A colonial research student at post-war II Cambridge', unpublished report, 26 September 2004.
15. Interview with Godfrey Stafford, 15 October 2004.
16. Lindsey G, 'A colonial research student at post-war II Cambridge', unpublished report, 26 September 2004.
17. Freeman J, *A Passion for Physics*, Institute of Physics Press, London, page 138, 1991.
18. Interview with John Wanklyn, 16 October 2004.
19. Wanklyn J, 'A Winter Holiday: Christmas 1947', unpublished report, page 1, 1 August 1998.
20. Interview with John Wanklyn, 16 October 2004.
21. Wanklyn J, 'A Winter Holiday: Christmas 1947', unpublished report, page 4, 1 August 1998.
22. Wanklyn J, 'A Winter Holiday: Christmas 1947', unpublished report,

page 5, 1 August 1998.

23. http://www.law.harvard.edu/library/about/news/2006/
SeaveyExhibition.php

24. Letter from Barbara Seavey to her parents, 22 September 1948.

25. Letter from Barbara Seavey to her parents, 8 October 1948.

26. Milburn RH, 'Excerpt: a Cambridge University Term', unpublished report, page 3, 4 January 2006.

27. Letter from Barbara Seavey to her parents, 8 October 1948.

28. http://www-groups.dcs.st-and.ac.uk/~history/
Mathematicians/Cartwright.html

29. Milburn RH, 'Excerpt: a Cambridge University Term', unpublished report, page 3, 4 January 2006.

30. Letter from Barbara Seavey to her parents, 25 January 1949.

31. Letter from Barbara Seavey to her parents, 6 November 1948.

32. Letter from Barbara Seavey to her parents, 12 December 1948.

33. Freeman J, *A Passion for Physics*, Institute of Physics Press, London, page 105, 1991.

34. Letter from Barbara Seavey to her parents, 23 October 1948.

35. Letter from Barbara Seavey to her parents, 27 December 1948.

36. Cormack AM, 'Autobiographical Sketch', unpublished report, Winchester, MA, page 5, 8 June 1990.

37. Letter from Barbara Seavey to her parents, 4 December 1948.

38. Letter from Barbara Seavey to her parents, 27 December 1948.

39. Letter from Barbara Seavey to her parents, 10 January 1949.

40. Letter from Barbara Seavey to her parents, 2 January 1949.

41. Letter from Barbara Seavey to her parents, 27 December 1948.

42. http://www.cies.org/about_fulb.htm

43. Letter from Barbara Seavey to her parents, 31 January 1949.

44. Letter from Barbara Seavey to her parents, 25 January 1949.

45. Letter from Barbara Seavey to her parents, 27 March 1949.

46. http://www.amherst.edu/~rjyanco94/literature/
rupertchawnerbrooke/poems/

47. Letter from Barbara Seavey to her parents, 8 March 1949.

48. Milburn RH, 'Excerpt: a Cambridge University Term', unpublished report, page 2, 4 January 2006.

49. Cormack AM, 'Autobiographical Sketch," unpublished report, Winchester, MA, page 5, 8 June 1990.

50. Freeman J, *A Passion for Physics*, Institute of Physics Press, London, page 137, 1991.

51. Letter from Barbara Seavey to her parents, 1 March 1949.

52. Letter from Barbara Seavey to her parents, 27 March 1949.

53. Letter from Barbara Seavey to her parents, 20 April 1949.

54. Lindsey G, 'A colonial research student at post-war II Cambridge', unpublished report, 26 September 2004.

55. Letter from Barbara Seavey to her parents, 5 July 1949.

56. Freeman J, *A Passion for Physics*, Institute of Physics Press, London, page 115, 1991.

57. Lindsey G, 'A colonial research student at post-war II Cambridge', unpublished report, 26 September 2004.

58. http://www.aip.org/history/lawrence/first.htm

59. http://www-outreach.phy.cam.ac.uk/camphys/

60. Pippard AB, 'The Cavendish Laboratory', *European Journal of Physics*, 8: 231–235, 1987.

61. Cormack AM, 'Autobiographical Sketch', unpublished report, Winchester, MA, page 5, 8 June 1990.

62. http://www-outreach.phy.cam.ac.uk/camphys/

63. Freeman J, *A Passion for Physics*, Institute of Physics Press, London, page 149, 1991.

64. Freeman J, *A Passion for Physics*, Institute of Physics Press, London, page 151, 1991.

65. Interview with Robin Cherry, 25 January 2005.

66. Freeman J, *A Passion for Physics*, Institute of Physics Press, London, 1991.

67. Letter from Brenton Stearns to Barbara Cormack, 14 March 1999.

68. Lindsey G, 'A colonial research student at post-war II Cambridge', unpublished report, 26 September 2004.

69. Cormack AM, 'Autobiographical Sketch', unpublished report, Winchester, MA, page 6, 8 June 1990.

70. Cormack AM, *Les Prix Nobel en 1979*, Nobel Foundation, Stockholm, page 169, 1980.

Chapter 4   Return to the Fairest Cape

1. Cormack AM, 'Recollections of my work with computer assisted tomography', *Tufts Criterion*, Volume XI, Number 6, page 9, November 1979.

2. Interview with Barbara Cormack, 3 October 2004.

3. Phillips H, *The University of Cape Town 1918–1948*, UCT Press, Cape

Town, page 301, 1993.

4. Allan Cormack's personal file, University of Cape Town, Memorandum of Agreement, 29 March 1950.

5. 'Universities Financially Embarrassed', *Varsity*, UCT Student Newsletter, page 3, 21 March 1951.

6. 'Holloway Commission Publishes its Report', *Varsity*, UCT Student Newsletter, page 1, 14 August 1953.

7. Cormack AM, 'Autobiographical Sketch', unpublished notes, Winchester, MA, page 6, 8 June 1990.

8. Letter from Allan Cormack to Professor RW James, 28 April 1951.

9. Louw JH, *In the Shadow of Table Mountain*, Struik Publishers, Cape Town, page 344, 1969.

10. Publications by the Department of Pathology in 1950, Annual Prospectus of the University of Cape Town, page 37, 1951.

11. Louw JH, *In the Shadow of Table Mountain*, Struik Publishers, Cape Town, page 399, 1969.

12. Letter from Allan Cormack to Professor EE Briggs of the Botany School at Cambridge University, 10 January 1951.

13. Polson A, 'Diffusion constants of the *E. Coli* bacteriophages', *Proceedings of the Society for Experimental Biology and Medicine*, 67: 294–296, 1948.

14. Louw JH, *In the Shadow of Table Mountain*, Struik Publishers, Cape Town, page 401, 1969.

15. Letter from Allan Cormack to Professor EE Briggs of the Botany School at Cambridge University, 10 January 1951.

16. Cover letter from Allan Cormack to Dr AFB Standfast, editor of the *Journal of General Microbiology*, 20 June 1952.

17. Letter from Allan Cormack to Professor Salvador Luria, 3 October 1962.

18. http://nobelprize.org/nobel_prizes/medicine/laureates/1969/luria-bio.html

19. Letter from Salvador Luria to Allan Cormack, 10 October 1962.

20. Editorial, 'Jubilee Science Congress, Cape Town, 1952', *South African Journal of Science*, 48(8): 265, 1952.

21. Editorial, *South African Journal of Science*, 49(3–4): 57–58, 1952.

22. E-mail correspondence with Colin Butler, 7 December 2006.

23. Schonland BFJ, 'The South African Association for the Advancement of Science, its past and future', *South African Journal of Science*, 49(3–4): 61–68, 1952.

24. Lemaître G, 'The clusters of nebulae in the expanding universe', *South African Journal of Science*, 49(3–4): 80–86, 1952.

25. Bragg L, 'Current researches in the Cavendish Laboratory', *South African Journal of Science*, 49(3-4): 109–114, 1952.

26. http://www.aps.org/publications/apsnews/200607/history.cfm

27. Bragg L, 'Current researches in the Cavendish Laboratory', *South African Journal of Science*, 49(3–4): 109–114, 1952.

28. Cormack AM, 'Autobiographical Sketch', unpublished notes, Winchester, MA, page 6, 8 June 1990.

29. Interview with Robin Cherry, 25 January 2005.

30. Cherry R, Obituary for Allan Cormack, *Sunday Times*, 24 May 1998.

31. Cormack AM, 'Momentum determination from cloud-chamber photographs in axially symmetric magnetic fields', *The Review of Scientific Instruments*, 24(3): 232, 1953.

32. Cormack AM, 'Resonance scattering of gamma rays by nuclei', *The Physical Review*, 96(3): 716–718, 1954.

33. Cormack AM, 'Neutrinos from the sun', *The Physical Review*, 97(1): 137–139, 1955.

34. http://www.zoominfo.com/search/PersonDetail.aspx?PersonID=152862855

35. Letter from Stanley Amdurer to Jean Cormack, 19 June 1998.

36. Teaching Staff, Annual Prospectus of the University of Cape Town, page 27, 1952.

37. Interview with Walter Powrie, 26 November 2006.

38. Letter from Colin 'Boot' Butler to Jean Cormack, 20 June 1998.

39. Letter to Allan Cormack from AVM Carter, Registrar of the University of Cape Town, 22 December 1951.

40. E-mail correspondence with Boot Butler, 7 December 2006.

41. Interview with Len Read, 20 July 2004.

42. Interview with Amy Read, 13 July 2004.

43. Interview with Barbara Cormack, 3 October 2004.

44. 'UCT Mourns its Chancellor', *Varsity*, UCT Student Newsletter, page 1, 20 September 1950.

45. 'No Progress Without Academic Freedom', *Varsity*, UCT Student Newsletter, page 1, 26 April 1951.

46. 'Debate on Colour Issue Indecisive', *Varsity*, UCT Student Newsletter, page 2, 11 June 1951.

47. 'Anti-UCT Campaign in Parliament and Press', *Varsity*, UCT Student Newsletter, page 1, 16 September 1953.

48. 'Quotable Quotes', *Varsity*, UCT Student Newsletter, page 2, 4 March 1954.
49. 'UCT's Liberty at Stake', *Varsity*, UCT Student Newsletter, page 1, March 1954.
50. Cormack AM, 'Autobiographical Sketch', unpublished notes, Winchester, MA, pages 6–7, 8 June 1990.
51. 'Chicago University Tells of Non-Segregation', *Varsity*, UCT Student Newsletter, page 4, 22 June 1954.
52. 'Die ANC en die Volkskongres', *Varsity*, UCT Student Newsletter, page 6, 16 September 1954.
53. Cherry RD, 'Professor RW James', *South African Journal of Science*, 93: 94–96, 1997.
54. 'Professor James Speaks Out', *Varsity*, UCT Student Newsletter, page 2, 26 March 1956.
55. Louw JH, *In the Shadow of Table Mountain*, Struik Publishers, Cape Town, page 383, 1969.
56. Prospectus of the University of Cape Town, Faculty of Medicine, page 45, 1955.
57. Cormack AM, 'The early developments of CT scan concepts', *Computed Tomographic Scanning of the Brain*, NIH Consensus Development Conference, Bethesda, MD, 5 November 1981.
58. Friedland GW, 'Memorial: Allan MacLeod Cormack, 1924–1998', *American Journal of Radiology*, 171: 1151–1152, 1998.
59. Videotape interview of Allan Cormack for TV programme *Naked to the Bone*, 11 March 1997.
60. Cormack AM, 'Chronology of my connection with computer assisted tomography', unpublished report, Tufts University, 10 February 1976.
61. Cormack AM, 'Recollections of my work with computer assisted tomography', *Tufts Criterion*, Volume XI, Number 6, page 9, November 1979.

Chapter 5   A New Beginning in Boston

1. Cormack AM, 'Autobiographical Sketch', unpublished notes, Winchester, MA, page 7, 8 June 1990.
2. Application for Furlough and/or Study Leave, Allan Cormack's personal file, University of Cape Town, 28 July 1955.
3. Special Exhibit on Professor Warren Seavey, Langdell Hall, Harvard University, 26 April 2006.

4. Interview with Dorothy and Bob Seavey, 23 April 2006.

5. Wilson R, *A Brief History of the Harvard University Cyclotrons*, Department of Physics, Harvard University Press, page 16, 2004.

6. Cormack AM, 'Autobiographical Sketch', unpublished notes, Winchester, MA, page 8, 8 June 1990.

7. 'Roy J Glauber', *Les Prix Nobel*. *The Nobel Prizes 2005*, edited by Karl Grandin, Nobel Foundation, Stockholm, 2006. http://nobelprize.org/nobel_prizes/physics/laureates/2005/glauber-autobio.html

8. Watson JD, Crick FHC, 'Molecular structure of nucleic acids', *Nature*, 171: 737–738, 1953.

9. Cormack AM, 'Fourier transforms in cylindrical co-ordinates', *Acta Crystallographica*, 10: 354–358, 1957.

10. Cormack AM, Presentation to American Psychiatry Association, Los Angeles, 6 May 1984.

11. Cormack AM, 'Report on Study Leave', Personal File, University of Cape Town, 3 September 1957.

12. Cormack AM, 'Autobiographical Sketch', unpublished notes, Winchester, MA, page 8, 8 June 1990.

13. Wilson R, *A Brief History of the Harvard University Cyclotrons*, Department of Physics, Harvard University Press, pp. 70–71, 2004.

14. Cormack AM, 'Autobiographical Sketch', unpublished notes, Winchester, MA, page 8, 8 June 1990.

15. 'Norman F Ramsey', *Les Prix Nobel*. *The Nobel Prizes 1989*, edited by Tore Frängsmyr, Nobel Foundation, Stockholm, 1990. http://nobelprize.org/nobel_prizes/physics/laureates/1989/ramsey-autobio.html

16. Cormack AM, 'Report on Study Leave', Personal File, University of Cape Town, 3 September 1957.

17. Palmieri JN, Cormack AM, Ramsey NF, Wilson R, 'Proton-proton scattering at energies from 46 to 147 MeV', *Annals of Physics*, 5(4): 299–341, 1958; and Cormack AM, Palmieri JN, Ramsey NF, Wilson R, 'Elastic scattering and polarization of protons by Helium at 147 and 66 MeV', *The Physical Review*, 115(3): 599–608, 1959.

18. Wilson R, *A Brief History of the Harvard University Cyclotrons*, Department of Physics, Harvard University Press, page 66, 2004.

19. Cormack AM, 'Autobiographical Sketch', unpublished notes, Winchester, MA, pp. 8–9, 8 June 1990.

20. Letter from Allan Cormack to Mr JG Benfield, Personal File, University

of Cape Town, 23 June 1957.

21. Letter from RW James to Allan Cormack, Personal File, University of Cape Town, 4 July 1957.

22. Cormack AM, 'Report on Study Leave', Personal File, University of Cape Town, 3 September 1957.

23. Interview with Robin Cherry, 25 January 2005.

24. Allan Cormack's Laboratory Notebook, Physics Department, University of Cape Town, 3 September 1957.

25. Friedland GW, 'Memorial: Allan MacLeod Cormack, 1924–1998', *American Journal of Radiology*, 171: 1151–1152, 1998.

26. Cormack AM, 'Reminiscences to Steve Webb, prompted by the publication *From the Watching of Shadows*', unpublished report, 15 August 1990.

27. Friedland GW, 'Memorial: Allan MacLeod Cormack, 1924–1998', *American Journal of Radiology*, 171: 1151–1152, 1998.

28. Letter from Professor DB Sears to Allan Cormack, 24 October 1957.

29. Letter from Allan Cormack to Professor EC Titchmarsh, 10 June 1959.

30. Letter from Professor EC Titchmarsh to Allan Cormack, 5 July 1959.

31. Cormack AM, 'Reminiscences to Steve Webb, prompted by the publication *From the Watching of Shadows*', unpublished report, page 2, 15 August 1990.

32. Friedland GW, 'Memorial: Allan MacLeod Cormack, 1924–1998', *American Journal of Radiology*, 171: 1151–1152, 1998.

33. E-mail correspondence with Brenton Stearns, 12 May 2005.

34. Ayres DS, Cormack AM, Greenberg AJ, Kenney RW, McLaughlin EF, Schafer RV, Caldwell DO, Elings VB, Hesse WP, Morrison RJ, 'An improved liquid hydrogen differential Ĉerenkov counter,' *Nuclear Instruments and Methods*, 70: 13–19, 1969.

35. Interview with Andreas Koehler, 4 October 2004.

36. Letter from Mildred Widgoff to Jean Cormack, 14 July 1998.

37. Wilson RR, 'Radiological use of fast protons', *Radiology*, 47: 487–491, 1946.

38. Wilson R, *A Brief History of the Harvard University Cyclotrons*, Department of Physics, Harvard University Press, page 9, 2004.

39. Kjellberg RN, Sweet WH, Preston WM, Koehler AM, 'The Bragg Peak of a proton beam in intracranial therapy of tumors', *Transactions of the American Neurological Association*, 87: 216–218, 1962.

40. E-mail correspondence with David Hennage, 22 January 2005.

41. Cormack AM, 'Representation of a function by its line integrals, with

some radiological applications', *Journal of Applied Physics*, 34(9): 2722–2727, 1963.

42. Webb S, *From the Watching of Shadows. The Origins of Radiological Tomography*, IOP Publishing, Adam Hilger, Bristol, page 256, 1990.

43. Brownell GL, Sweet WH, 'Localisation of brain tumours with positron emitters', *Nucleonics*, 11: 40–45, 1953.

44. Schneider RJ, Cormack AM, 'Neutron total cross sections in the energy range 100 to 150 MeV', *Nuclear Physics*, A119: 197–208, 1968.

45. Cormack AM, 'Reminiscences to Steve Webb, prompted by the publication *From the Watching of Shadows*', unpublished report, page 3, 15 August 1990.

46. E-mail correspondence with David Hennage, 22 January 2005.

47. Letter from Robert J Schneider to Jean Cormack, 3 June 1998.

48. Cormack AM, 'Reminiscences to Steve Webb, prompted by the publication *From the Watching of Shadows*', unpublished report, page 4, 15 August 1990.

49. Cormack AM, 'Record of appointment to see Brownell and Aronow about the results in the '64 paper', photocopy of personal diary, 16 to 22 December 1963.

50. Wilson R, *A Brief History of the Harvard University Cyclotrons*, Department of Physics, Harvard University Press, page 94, 2004.

51. Cormack AM, 'Proposal for the support of some experiments pertinent to the work of the Medical Group at the Harvard Cyclotron', grant application to NASA, 1964.

52. Cormack AM, 'Reminiscences to Steve Webb, prompted by the publication *From the Watching of Shadows*', unpublished report, page 5, 15 August 1990.

53. Cormack AM, 'Representation of a function by its line integrals, with some radiological applications. II', *Journal of Applied Physics*, 35(10): 2908–2913, 1964.

54. E-mail correspondence with Margaret Cormack, 7 March 2005.

55. Interview with Jean Cormack, 23 April 2006.

56. 'The Town of Alton, New Hampshire 1770–2004', http://www.alton.nh.gov/

57. Interview with Dorothy Seavey, 23 April 2006.

58. Interview with Margaret Cormack, 28 April 2006.

59. Curriculum vitae of Barbara J Cormack, October 2004.

60. E-mail correspondence with Margaret Cormack, 10 June 2006.

61. Chaikin A, 'Greatest space events of the 20th Century: the 60s', *Space*

and Science, 27 December 1999, http://www.space.com/

62. Letters from Senator John Glenn to Allan Cormack dated 5 May and 30 July 1982, and letter from Allan Cormack to Senator Glenn dated 25 May 1982.

Chapter 6   Finding Radon and his Transform

1. Cormack AM, 'Reminiscences to Steve Webb, prompted by the publication *From the Watching of Shadows*', unpublished report, page 6, 15 August 1990.
2. Letter from Allan Cormack to Professor Claude Jaccard, 5 November 1980.
3. Cormack AM, 'Early two-dimensional reconstruction and recent topics stemming from it', *Medical Physics*, 7(4): 277–282, 1980.
4. Letter from Claude Jaccard to Allan Cormack, 5 December 1980.
5. Jaccard C, 'Automatic thin section analysis with the tomography', *Mitteilungen des Eidgenössischen Institut für Schnee-und Lawinenforschung*, Davos, No. 28, pp. 345–351, 1974.
6. Videotape interview of Allan Cormack for programme *Naked to the Bone*, 11 March 1997.
7. Schneider RJ, Cormack AM, 'Neutron total cross sections in the energy range 100 to 150 MeV', *Nuclear Physics*, A119: 197–208, 1968.
8. Letter from Allan Cormack to Mildred Widgoff, 11 May 1965.
9. Wilson R, *A Brief History of the Harvard University Cyclotrons*, Department of Physics, Harvard University Press, page 84, 2004.
10. Horvitz B, 'Explosion rocks Cambridge Electron Accelerator', *The Tech*, page 11, 22 September 1965.
11. Interview with Andy Koehler, 4 October 2004.
12. Letter from Allan Cormack to Mildred Widgoff, 30 July 1965.
13. Wilson R, *A Brief History of the Harvard University Cyclotrons*, Department of Physics, Harvard University Press, page 84, 2004.
14. Interview with Andy Koehler, 4 October 2004.
15. Horvitz B, 'Explosion rocks Cambridge Electron Accelerator', *The Tech*, page 11, 22 September 1965.
16. Interview with Margaret Cormack, 28 April 2006.
17. E-mail correspondence with Brenton Stearns, 12 May 2005.
18. Interview with Jean Cormack, 23 April 2006.
19. Interview with Margaret Cormack, 28 April 2006.
20. http://en.wikipedia.org/wiki/Lawrence_Berkeley_National_

Laboratory

21. E-mail correspondence with David Ayres, 3 March 2007.

22. E-mail correspondence with Brenton Stearns, 12 May 2005

23. Ayres DS, Cormack AM, Greenberg AJ, Kenney RW, Caldwell DO, Elings VB, Hesse WP, Morrison RJ, 'Precise comparison of π^+ and π^- lifetimes', *Physical Review Letters*, 21(4): 261–265, 1968.

24. E-mail correspondence with David Ayres, 3 March 2007.

25. Ayres DS, Cormack AM, Greenberg AJ, Kenney RW, Caldwell DO, Elings VB, Hesse WP, Morrison RJ, 'A determination of the refractive index of liquid deuterium using the Čerenkov effect', *Physica*, 43: 105–108, 1969.

26. Interview with Margaret Cormack, 28 April 2006.

27. E-mail correspondence with Margaret Cormack, 30 April 2005.

28. Interview with Kathryn McCarthy, 23 April 2006.

29. Letter from Amelia Cormack to Allan Cormack, 3 April 1967.

30. Cormack AM, 'Reminiscences to Steve Webb, prompted by the publication *From the Watching of Shadows*', unpublished report, 15 August 1990.

31. John F, *Plane Waves and Spherical Means Applied to Partial Differential Equations*, Wiley Interscience, New York, 1955.

32. Radon J, 'Über die Bestimmung von Funktionen durch ihre Integralwerte längs gewisser Mannigfaltigkeiten', *Berichte über die Verhandlungen der Königlich Sächsichen Gesellschaft der Wissenschaften zu Leipzig*, BG Teubner, Leipzig, 69: 262–277, 1917.

33. Interview with Margaret Cormack, 28 April 2006.

34. Bukovics B, 'Biography of Johann Radon', Vienna, Austria, August 1992.

35. Cormack AM, 'Reminiscences to Steve Webb, prompted by the publication *From the Watching of Shadows*', unpublished report, 15 August 1990.

36. E-mail correspondence with Michael Goitein, 9 April 2005.

37. Webb S, *From the Watching of Shadows. The Origins of Radiological Tomography*, IOP Publishing, Adam Hilger, Bristol, page 188, 1990.

38. Letter from Allan Cormack to Michael Goitein, 15 June 1971.

39. Letter from Michael Goitein to Allan Cormack, 16 February 1972.

40. Goitein M, 'Three-dimensional density reconstruction from a series of two-dimensional projections', *Nuclear Instruments and Methods*, 101: 509–518, 1972.

41. Crowther RA, De Rosier DJ, Klug A, 'The reconstruction of a three-

dimensional structure from projections and its application to electron microscopy', *Proceedings of the Royal Society of London*, 317: 319–340, 1970.

42. Letter from Allan Cormack to Aaron Klug, 15 December 1971.

43. Letter from Aaron Klug to Allan Cormack, 27 January 1972.

44. Letter from John Hale, editor of *Physics in Medicine and Biology*, to Allan Cormack, 19 July 1972.

45. Cormack AM, 'Reconstruction of densities from their projections, with applications in radiological physics', *Physics in Medicine and Biology*, 18(2): 195–207, 1973.

46. Ambrose J, Hounsfield G, 'Computerized transverse axial tomography', *British Journal of Radiology*, 46: 148–149, 1973.

47. E-mail correspondence with Michael Goitein, 9 April 2005.

48. Friedland GW, Thurber BD, 'The birth of CT', *American Journal of Radiology*, 167: 1365–1370, 1996.

49. Cormack AM, 'Reminiscences to Steve Webb, prompted by the publication *From the Watching of Shadows*', unpublished report, page 7, 15 August 1990.

50. Webb S, *From the Watching of Shadows. The Origins of Radiological Tomography*, IOP Publishing, Adam Hilger, Bristol, page 192, 1990.

51. Petrik V, Apok V, Britton JA, Bell BA, Papadopoulos MC, 'Godfrey Hounsfield and the dawn of computed tomography', *Neurosurgery*, 58: 780–787, 2006.

52. Letter from Steve Webb to Allan Cormack, 22 August 1990.

53. Letter from Allan Cormack to Gerald Friedland, 15 May 1996.

54. Petrik V, Apok V, Britton JA, Bell BA, Papadopoulos MC, 'Godfrey Hounsfield and the dawn of computed tomography', *Neurosurgery*, 58: 780–787, 2006.

55. Friedland GW, Thurber BD, 'The birth of CT', *American Journal of Radiology*, 167: 1365–1370, 1996.

56. Hounsfield GN, 'Computerized transverse axial scanning (tomography): Part I. Description of system', *British Journal of Radiology*, 46: 1016–1022, 1973.

57. Friedland GW, Thurber BD, 'The birth of CT', *American Journal of Radiology*, 167: 1365–1370, 1996.

58. Kevles BH, *Naked to the Bone. Medical Imaging in the Twentieth Century*, Addison-Wesley, Reading, Massachusetts, page 162, 1997.

59. Gordon R, Bender R, Herman GT, 'Algebraic reconstruction techniques (ART) for three-dimensional electron microscopy and x-ray photogra-

phy', *Journal of Theoretical Biology*, 29: 471–482, 1970.

60. De Rosier DJ, Klug A, 'Reconstruction of three dimensional structures from electron micrographs', *Nature*, 217: 130–134, 1968.

61. Crowther RA, Klug A, 'ART and science or conditions for three-dimensional reconstruction from electron microscope images', *Journal of Theoretical Biology*, 32: 199–203, 1971.

62. Letter from Aaron Klug to Allan Cormack, 27 January 1972.

63. E-mail correspondence with Bob Marr, 16 May 2005.

64. Gordon R, Herman GT, '3-dimensional reconstruction from projections: review of algorithms', *International Review of Cytology - A Survey of Cell Biology*, 38: 111–151, 1974.

65. Marr RB, Editor of *Techniques of Three-Dimensional Reconstruction*, Proceedings of an International Workshop held at Brookhaven National Laboratory, Upton, New York, 166 pages, 1974.

66. Bracewell RN, 'Strip integration in radio-astronomy', *Australian Journal of Physics*, 9: 198–217, 1956.

67. Cormack AM, 'Reminiscences to Steve Webb, prompted by the publication *From the Watching of Shadows*', unpublished report, page 7, 15 August 1990.

68. E-mail correspondence with Paul Lauterbur, 22 November 2004.

69. Koehler AM, 'Proton radiography', *Science*, 160(3825): 303–304, 1968.

70. Wilson R, *A Brief History of the Harvard University Cyclotrons*, Department of Physics, Harvard University Press, page 35, 2004.

71. Steward VW, Koehler AM, 'Proton radiographic detection of strokes', *Nature*, 245(5419): 38–40, 1973.

72. Steward VW, Koehler AM, 'Proton radiography in the diagnosis of breast carcinoma', *Radiology*, 110(1): 217–221, 1974.

73. Letter from Allan Cormack to Bill Steward, 30 November 1973.

74. Letter from Rodney Brooks to Allan Cormack, 23 January 1975.

75. Ledley RS, Di Chiro G, Luessenhop AJ, Twigg HL, 'Computerized transaxial X-ray tomography of the human body', *Science*, 186(4160): 207–212, 1974.

76. Letter from Rodney Brooks to Allan Cormack, 19 February 1975.

77. Cormack AM, Koehler AM, 'Quantitative proton tomography: preliminary experiments', *Physics in Medicine and Biology*, 21(4): 560–569, 1976.

78. E-mail correspondence with Rodney Brooks, 25 March 2005.

79. Interview with Dorothy Seavey, 23 April 2006.

80. Interview with Robert Cormack, 23 April 2006.

81. Letter from Allan Cormack to Godfrey Stafford, 20 October 1975.
82. Letter from Allan Cormack to Bernard Gordon, 24 October 1975.
83. Tsuruoka D, 'Physicist Allan Cormack changed medicine by taking simple approach', *Investor's Business Daily*, 27 April 2001.
84. Gordon BM, Neumann L, Dobbs JM, 'Tomography signal processing system', United States Patent Number 4,135,247, 16 January 1979.
85. Letter from Giovanni Di Chiro to Allan Cormack, 22 October 1975.

Chapter 7   On the Road to Stockholm

1. Acceptance speech by Allan Cormack, Nobel Award Ceremony, Stockholm, Sweden, 10 December 1979.
2. Ambrose J, 'A brief review of the EMI scanner', *British Journal of Radiology*, 48(571): 605–606, 1975.
3. Letter from Allan Cormack to the editor of the *British Journal of Radiology*, entitled 'More on the EMI scanner', 20 November 1975.
4. Letter from Mrs L Surry to Allan Cormack, 10 December 1975.
5. Letter from Allan Cormack to James Ambrose, 16 December 1975.
6. Letter from James Ambrose to Allan Cormack, 8 January 1976.
7. Letter from Dr DK Bewley to Allan Cormack, 15 January 1976.
8. Letter from Allan Cormack to Robin Cherry and John Juritz, 1 December 1975.
9. Letter from Robin Cherry to Allan Cormack, 19 December 1975.
10. Cormack AM, 'Chronology of my Connection with Computer Assisted Tomography', unpublished report, Tufts University, 10 February 1976.
11. Letter from Burton Hallowell to Salvatore Luria, 23 February 1976.
12. Cormack AM, Koehler AM, Brooks RA, Di Chiro G, 'Proton tomography: progress below the 0.5 % density difference level', *Proceedings of International Symposium and Course of Computerized Tomography*, edited by Juan Taveras and Paul Frew, San Juan, Puerto Rico, 5 to 9 April 1976.
13. Letter from Allan Cormack to Torgny Greitz, 28 April 1976.
14. Letters from Allan Cormack to Torgny Greitz and Björn Nordenström, 7 June 1976.
15. Brooks RA, Di Chiro G, 'Principles of computer assisted tomography (CAT) in radiographic and radioisotopic imaging', *Physics in Medicine and Biology*, 21(5): 689–732, 1976.
16. Interview with Jean Cormack, 23 April 2006.
17. Interview with Robert Cormack, 23 April 2006.

18. Interview with Margaret Cormack, 28 April 2006.

19. Cormack AM, 'Some family history', unpublished report, Winchester, MA, 11 pages, Winchester, Massachusetts, 1996.

20. Cormack AM, submission to *Science at the Bicentennial. A Report from the Research Community*, 8th Annual Report of the National Science Board to the President, pages 76–77, 1976.

21. E-mail correspondence with Margaret Cormack, 30 April 2005.

22. Cormack AM, Doyle BJ, 'Algorithms for two-dimensional reconstruction', *Physics in Medicine and Biology*, 22(5): 994–997, 1977.

23. Crowther RA, Klug A, Hounsfield GN, Isherwood I, Pullan BR, Rutherford RA, 'Three-dimensional image reconstruction in electron microscopy and X-radiography', *Conversazione*, Royal Society, London, November 1975.

24. Letter from Aaron Klug to Allan Cormack, 10 September 1976.

25. Letter from Ron Bracewell to Allan Cormack, 30 September 1976.

26. 'Investigators hail use of CAT scanners as great advance in X-ray diagnosis', *The NIH Record*, 2 November 1976.

27. Di Chiro G, Editor of *International Symposium on Computer Assisted Tomography in Nontumoral Diseases of the Brain, Spinal Cord and Eye*, National Institutes of Health, Bethesda, Maryland, page 39, October 1976.

28. E-mail correspondence with David Kuhl, 25 March 2005.

29. Oldendorf WH, 'Isolated flying spot detection of radiodensity discontinuities; displaying the internal structural pattern of a complex object', *IRE Transactions on Biomedical Electronics*, BME-8: 68–72, 1961.

30. Unpublished letter from Allan Cormack to *The New York Times*, following an obituary for William Oldendorf, 5 January 1993.

31. Cormack AM, Koehler AM, 'The use of high energy protons in the non-destructive examination of materials', *Proceedings of International Symposium on Archaeometry and Archaeological Prospection*, edited by R Maddin, University of Pennsylvania, Philadelphia, page 59, March 1977.

32. Letter from Allan Cormack to Robert Maddin, 13 January 1977.

33. Letter from John Slocum to Allan Cormack, 17 May 1977.

34. Letter from Allan Cormack to John Slocum, 27 May 1977.

35. Letter from Allan Cormack to Philip DuBois, 21 April 1978.

36. Letter from Richard Wilson to Philip DuBois, 25 January 1978.

37. Letter from Allan Cormack to Philip DuBois, 22 March 1978.

38. Letter from Philip DuBois to Allan Cormack, 12 August 1978.

39. Martin RL, 'Development of a prototype proton CAT scan system', Grant application to the National Cancer Institute, N01-CB-43918, 2 November 1977.

40. Kramer SL, Moffett DR, Martin RL, Colton EP, Steward VW, 'Proton imaging for medical applications', *Radiology*, 135(2): 485–494, 1980.

41. Cormack AM, 'Letter to the editor', *Physics Bulletin*, 28: 543, 1977.

42. E-mail correspondence with Rodney Brooks, 25 March 2005.

43. Hounsfield GN, 'Method and apparatus for measuring X or gamma radiation absorption or transmission at plural angles and analyzing the data', United States Patent Number 3,778,614, 11 December 1973.

44. Webb S, *From the Watching of Shadows. The Origins of Radiological Tomography*, IOP Publishing, Adam Hilger, Bristol, page 196, 1990.

45. Strong AB, Hurst RAA, 'EMI patents on computed tomography: history of legal actions', *British Journal of Radiology*, 67(795): 315–316, 1994.

46. Broad WJ, 'Riddle of the Nobel debate', *Science*, 207: 37–38, 1980.

47. Cormack AM, 'Reminiscences to Steve Webb, prompted by the publication *From the Watching of Shadows*', unpublished report, 15 August 1990.

48. Letter from AB Logan to Denis Rutovitz, 23 June 1981.

49. Letter from Denis Rutovitz to Godfrey Hounsfield, 9 June 1981.

50. Broad WJ, 'Riddle of the Nobel debate', *Science*, 207: 37–38, 1980.

51. Letter from Allan Cormack to Gerald Friedland, 15 May 1996.

52. Cormack AM, 'EMI patent litigation in the US', *British Journal of Radiology*, 67(795): 316–317, 1994.

53. Letter from Allan Cormack to Fearghus O'Foghludha, 24 March 1977.

54. Cormack AM, Doyle BJ, 'Algorithms for two-dimensional reconstruction', *Physics in Medicine and Biology*, 22(5): 994–997, 1977.

55. Letter from PR Moran to Allan Cormack, 15 July 1980.

56. E-mail correspondence with Robert Lewitt, 24 May 2006.

57. Bates RHT, Peters TM, 'Towards improvements in tomography', *New Zealand Journal of Science*, 14: 883–896, 1971.

58. Webb S, *From the Watching of Shadows. The Origins of Radiological Tomography*, IOP Publishing, Adam Hilger, Bristol, page 187, 1990.

59. Letters from Robert Lewitt to Allan Cormack, dated 7 September, 3 November, 2 December, 12 December, 14 December, and 30 December 1977.

60. Letter from Allan Cormack to Robert Lewitt, 9 December 1977.

61. Letter from Allan Cormack to Brian Doyle, 9 December 1977.

62. Interview with Todd Quinto, 6 October 2004.

63. Quinto ET, 'Report on the Hole Theorem', unpublished report, Tufts University, 12 January 1979.

64. Letter from Robert Lewitt to the British editor of *Physics in Medicine and Biology*, 31 August 1979.

65. Letter from Allan Cormack to the British editor of *Physics in Medicine and Biology*, 19 June 1979.

66. E-mail correspondence with Marcella Tanona, 16 February 2007.

67. Interview with Kathryn McCarthy, 23 April 2006.

68. Kember NF, 'Social irresponsibility?', *Physics Bulletin*, Volume 28, November 1977.

69. Pippard AB, 'Social irresponsibility?', *Physics Bulletin*, Volume 28, October 1977.

70. Letter from Allan Cormack to the editor of *Physics Bulletin*, 4 November 1977.

71. Barbara Cormack's curriculum vitae, October 2004.

72. Letter from Allan Cormack to John and June Juritz, 26 January 1980.

73. Interview with Margaret Cormack, 28 April 2006.

74. Cormack AM, *Les Prix Nobel en 1979*, Nobel Foundation, Stockholm, page 170, 1980.

75. Phillips H, *The University of Cape Town 1918-1948. The Formative Years*, UCT Press, Cape Town, page 354, 1993.

76. Douglas E, 'Darwin's natural heir', *The Guardian*, 17 February 2001.

77. Wilson EO, *Consilience: The Unity of Knowledge*, Alfred A Knopf, New York, page 6, 1998.

78. Interview with Jean Cormack, 23 April 2006.

79. Cooke R, 'To Cormack, a sideline', *Boston Globe*, 12 October 1979.

80. Blum L, 'Cormack besieged after Nobel Prize, to continue coursework on the hill', *Tufts Criterion*, 19 October 1979.

81. Sanderson B, 'Cormack awarded Nobel Prize', *The Tufts Observer*, Volume 14, Number 6, 12 October 1979.

82. Blum L, 'New policy spurs divestment of 2 companies', *The Tufts Observer*, Volume 14, Number 6, 12 October 1979.

83. Letter from Stig Ramel to Allan Cormack, 11 October 1979.

84. Letters from Senators Paul Tsongas and Edward Kennedy to Allan Cormack dated 11 and 12 October 1979, respectively.

85. Letter from President Jimmy Carter to Allan Cormack, 12 October 1979.

86. Broad WJ, 'Riddle of the Nobel debate', *Science*, 207: 37-38, 1980.

87. Unpublished letter from Allan Cormack to *The New York Times*, following the death of William Oldendorf, 5 January 1993.

88. Webb S, *From the Watching of Shadows. The Origins of Radiological Tomography*, IOP Publishing, Adam Hilger, Bristol, page 183, 1990.

89. Letter from Ulf Rudhe to Allan Cormack, 16 February 1981.

90. Barrett HH, Hawkins WG, Joy MLG, 'Historical note on computed tomography', *Radiology*, 147: 172, 1983.

91. Vaughan CL, Mayosi BM, 'Origins of computed tomography', *The Lancet*, 369: 1168, 7 April 2007.

92. Letter from John Clifton to Allan Cormack, 12 October 1979.

93. Letter from John Clifton to Robert Lewitt, 3 December 1979.

94. Cormack AM, 'Algorithms for two-dimensional reconstruction' (letter), *Physics in Medicine and Biology*, 25(2): 372, 1980.

95. 'The Nobel Prizes: That Winning American Style', *Time Magazine*, pages 66–68, 29 October 1979.

96. E-mail correspondence with Margaret Cormack, 30 April 2005.

97. Di Chiro G, Brooks RA, 'The 1979 Nobel Prize in Physiology or Medicine', *Science*, 206: 1060–1062, 1979.

98. Letter from Kenneth Hanson to Allan Cormack, 16 November 1979.

99. Cormack AM, 'Reminiscences to Steve Webb, prompted by the publication *From the Watching of Shadows*', unpublished report, page 8, 15 August 1990.

100. Friedland GW, 'Allan MacLeod Cormack, 1924–1998', *American Journal of Radiology*, 171: 1151–1152, 1998.

101. Interview with Margaret Cormack, 28 April 2006.

102. Interview with Jean Cormack, 23 April 2006.

103. Letter from Allan Cormack to John and June Juritz, 26 January 1980.

104. E-mail correspondence with Margaret Cormack, 30 April 2005.

105. Hounsfield GN, 'Computed medical imaging', *Medical Physics*, 7(4): 283–290, 1980.

106. Interview with Robert Cormack, 23 April 2006

107. E-mail correspondence with Jean Cormack, 11 March 2005.

108. E-mail correspondence with Margaret Cormack, 30 April 2005.

109. Letter from Allan Cormack to John and June Juritz, 26 January 1980.

110. *Les Prix Nobel 1979*, Nobel Foundation, Stockholm, page 41, 1980.

111. Friedland GW, 'Allan MacLeod Cormack, 1924–1998', *American Journal of Radiology*, 171: 1151–1152, 1998.

112. Letter from Allan Cormack to John and June Juritz, 26 January 1980.

Chapter 8   Citizen of the World

1. Letter from Allan Cormack to Gary Barnes, President of the American Association of Physicists in Medicine, 15 December 1988.
2. Interview with Margaret Cormack, 28 April 2006.
3. Citation by Tufts University when conferring the honorary Doctor of Science degree on Allan Cormack, 25 May 1980.
4. 'UCT gold for Nobel Prize man', *The Argus*, 15 July 1980.
5. 'Science – facts but no humanity?' *Cape Times*, 15 July 1980.
6. Versveld M, 'Bomb Condemned', *Varsity*, UCT Student Newsletter, page 3, 27 August 1945.
7. 'A hobby won UCT man the Nobel Prize', *The Argus*, 15 July 1980.
8. 'Cormack lecture on Monday', *Cape Times*, 14 March 1981.
9. 'Nobel Centenary: stamps honour South Africa's laureates', *Setempe*, South African Post Office, pages 19 to 21, 4 November 1996.
10. 'National medals are pinned on 30 scientists', *Washington Post*, 15 November 1990.
11. Saltus R, 'Seven area researchers win science awards', *Boston Globe*, 14 November 1990.
12. Letter from Allan Cormack to Richard Weiss, chairman of the Tufts mathematics department, August 1983.
13. Letter from Todd Quinto to the Cormack family, 9 May 1998.
14. Interview with Todd Quinto, 6 October 2004.
15. Eulogy by Todd Quinto at the memorial service for Allan Cormack, 30 June 1998.
16. Cormack AM, Quinto ET, 'A Radon transform on spheres through the origin in R^n and applications to the Darboux equation', *Transactions of the American Mathematical Society*, 260(2): 575–581, 1980.
17. Letter from Todd Quinto to Jean Cormack, 14 August 1998.
18. Cormack AM, 'Early tomography and related topics', in *Mathematical Aspects of Computerized Tomography*, edited by GT Herman and F Natterer, Springer-Verlag, New York, pp. 1–6, 1981.
19. Letter from Allan Cormack to Richard Weiss, chairman of the Tufts mathematics department, August 1983.
20. Cormack AM, Quinto ET, 'On a problem in radiotherapy: questions of non-negativity', *International Journal of Imaging Systems and Technology*, 1: 120–124, 1989. Cormack AM, Quinto ET, 'The mathematics and physics of radiation dose planning using X-rays', *Contemporary Mathematics*, 113: 41–55, 1990.

21. Eulogy by Todd Quinto at the memorial service for Allan Cormack, 30 June 1998.

22. Brahme A, Roos JE, Lax I, 'Solution of an integral equation encountered in rotation therapy', *Physics in Medicine and Biology*, 27: 1221–1229, 1982.

23. Cormack AM, 'A problem in rotation therapy with X rays', *International Journal of Radiation Oncology, Biology, Physics*, 13(4): 623–630, 1987.

24. Cormack AM, Cormack RA, 'A problem in rotation therapy with X-rays: dose distributions with an axis of symmetry', *International Journal of Radiation Oncology, Biology, Physics*, 13(12): 1921–1925, 1987.

25. Goitein M, 'The inverse problem', *International Journal of Radiation Oncology, Biology, Physics*, 18: 489–491, 1990.

26. Cormack AM, 'Response to Goitein', *International Journal of Radiation Oncology, Biology, Physics*, 18: 709, 1990.

27. E-mail correspondence with Michael Goitein, 9 April 2005.

28. Wilson R, *A Brief History of the Harvard University Cyclotrons*, Department of Physics, Harvard University Press, page 45, 2004.

29. Letter from Herman Suit to Allan Cormack, 26 June 1995.

30. Cormack AM, 'Behind the great leap forward', paper presented at Proton Therapy Symposium, Massachusetts General Hospital, 8 pages, 14 September 1995.

31. E-mail correspondence with Michael Goitein, 9 April 2005.

32. Grünbaum FA, 'Reconstruction with arbitrary directions: dimensions two and three', *Mathematical Aspects of Computerized Tomography*, edited by GT Herman and F Natterer, Springer-Verlag, New York, pp. 112–126, 1981.

33. Bockwinkel HBA, 'Over de voortplanting van licht in een twee-assig kristal rondom een middelpunt van trilling', *Verh. Konink. Acad. V. Wet. Wissen. Natur.*, 14(2): 636–651, 1906.

34. Uhlenbeck GE, 'Over een stelling van Lorentz en haar uitbreiding voor meerdimensionale ruimten', *Physica, Nederlandsch Tijdschrift voor Natuurkunde*, 5: 423–428, 1925.

35. Knipp JK, Uhlenbeck GE, 'Emission of gamma radiation during the beta decay of nuclei', *Physica*, 3(6): 425–439, 1936.

36. Cormack AM, 'Internal Rayleigh scattering', *The Physical Review*, 97(4): 986, 1955.

37. Letter from Allan Cormack to Dr AJ Kox, April 1981.

38. Letter from Dr AJ Kox to Allan Cormack, 21 May 1981.

39. Cormack AM, '75 years of Radon transform', *Journal of Computer Assisted Tomography*, 16(5): 673, 1992.
40. Letter from Viktor Ambartsumian to Allan Cormack, 6 April 1981.
41. Ambartzumian V, 'On the derivation of the frequency function of space velocities of the stars from the observed radial velocities', *Monthly Notices of the Royal Astronomical Society*, 96(3): 172–178, 1936.
42. Letter from Allan Cormack to Viktor Ambartsumian, 9 June 1981.
43. Cormack AM, 'On the future of the principles of computed tomography', *Toshiba Medical Review*, 1981.
44. Letter from Allan Cormack to Dr Brigitte Bukovics, daughter of Johann Radon, 25 May 1993.
45. Cormack AM, 'My connection with the Radon transform', *Proceedings of the Conference 75 Years of the Radon Transform*, International Press, pp. 32–35, 1994.
46. Cormack AM, 'A paraboloidal Radon transform', *Proceedings of the Conference 75 Years of the Radon Transform*, International Press, pp. 105–109, 1994.
47. Letter from Takayoshi Matsui to Allan Cormack, 5 February 1977.
48. Interview with Margaret Cormack, 28 April 2006.
49. Letter from Allan Cormack to Takayoshi Matsui, January 1981.
50. Letter from Daisuke Yamauchi to Allan Cormack, 6 January 1984.
51. Cormack AM, 'Technical and moral problems for the 21st century', *Shueisha Imidas International Seminar*, Tokyo, Japan, 9 March 1988.
52. Letter from Hiroki Goto to Allan Cormack, 10 January 1992.
53. 'Dr Allan Cormack: Nobel Prize Winner in Physiology or Medicine in 1979', (in Japanese), *The Weekly Shonen Jump*, 33: 249–255, 1992.
54. Letter from Fukushima Seiji of Say International to Allan Cormack, 12 January 1994.
55. http://www.lindau-nobel.de/
56. '23 Laureaten aus Europa und der USA kommen nach Lindau', *Lindauer Zeitung*, 23 June 1984.
57. Broad WJ, 'Riddle of the Nobel debate', *Science*, 207: 37–38, 1980.
58. E-mail correspondence with Margaret Cormack, 30 April 2005.
59. Eccles JC, 'Do mental events cause neural events analogously to the probability fields of quantum mechanics?', *Proceedings of the Royal Society of London*, B 227: 411–428, 1986.
60. Letter from Countess Sonja Bernadotte to Allan Cormack, 2 August 1989.
61. Letter from Allan Cormack to John Wanklyn, 25 July 1996.

62. Cormack AM, 'Stockholm 1991', unpublished report, Winchester, MA, page 1, 14 January 1992.

63. Interview with Aaron Klug, 18 October 2004.

64. Brahme A, 'Similarities and differences between computed tomography and radiation therapy planning', *Specialla Arrangemang för Tidigare Nobelpristagare*, Radiumhemmet, Karolinska Sjukhuset, 7 December 1991.

65. Cormack AM, 'Stockholm 1991', unpublished report, Winchester, MA, page 5, 14 January 1992.

66. Vaughan CL, 'Digital X-rays come of age', *South African Medical Journal*, 96(7):610–612, 2006.

67. Röntgen WC, 'Über eine neue Art von Strahlen', *Sitzungs-Berichte der physik.-med. Gesellschaft zu Würzburg*, 137: 132–141, 1895.

68. Cormack AM, 'South Korea 1995', unpublished report, Winchester, MA, 4 June 1995.

69. Letter from David Moncton to Allan Cormack, 20 September 1995.

70. Cormack AM, 'Röntgen celebrations and related matters', unpublished report, Winchester, MA, 4 pages, 28 November 1995.

71. Letter from Axel Haase to Allan Cormack, 31 March 1995.

72. Letter from Allan Cormack to Axel Haase, 13 November 1995.

73. Letter from Allan Cormack to Robert Adams, offering to donate his prototype CAT-scanner to the Smithsonian Institution, 22 March 1989.

Chapter 9   At Home in Massachusetts

1. Letter from Allan Cormack to Andrew Kadak, 31 August 1989.

2. http://www.jsc.nasa.gov/Bios/htmlbios/hauck-fh.html

3. Letter from Frederick Hauck to Allan Cormack, 20 November 1980.

4. Letter from Ross Humphreys to Allan Cormack, 16 December 1980.

5. Di Chiro G, Brooks RA, 'The 1979 Nobel Prize in Physiology or Medicine', *Science*, 206: 1060–1062, 30 November 1979.

6. Letter from William Ross Humphreys to Nancy Seymour of Tufts University, 15 December 1980.

7. Letter from Allan Cormack to Ross Humphreys, 2 January 1981.

8. http://www.stanfordalumni.org/news/magazine/2000/mayjun/upfront/whos_who.html

9. http://www.gbpatent.com/professionals.html

10. Letter from Allan Cormack to Bruce Bernstein, 21 June 1983.

11. E-mail correspondence with Matt Howard, 6 April 2005.

12. Letter from Matt Howard to Todd Quinto, 19 October 1989.

13. Howard MA, Grady MS, Ritter RC, 'A magnetic surgery system', United States Patent Number 4,869,247, issued 26 September 1989.

14. Letter from Rogers Ritter to Allan Cormack, 5 December 1989.

15. http://www.stereotaxis.com/

16. Howard MA, Bell BA, Uttley D, 'The pathophysiology of infant subdural haematomas', *British Journal of Neurosurgery*, 7(4): 355–365, 1993.

17. 'The nuclear agenda', editorial in the *Worcester Telegram*, 17 August 1988.

18. Letter from Robert Palmer, campaign coordinator for 'Massachusetts Citizens Against the Shutdown Initiative', to Allan Cormack, 1 July 1988.

19. Letter from Jerome Wiesner, President Emeritus of MIT, to 'Scientists and Engineers for Secure Energy', 2 March 1988.

20. 'The greenhouse referendum', editorial in *The Boston Globe*, 15 August 1988.

21. Memorandum from Robert Palmer to the committee members of 'Massachusetts Citizens Against the Shutdown Initiative', 2 September 1988.

22. Letter from Stephen Sweeney, chief executive officer of Boston Edison, to Allan Cormack, 8 December 1988.

23. Letter from Andrew Kadak to Allan Cormack, 20 January 1989.

24. Letter from Allan Cormack to Andrew Kadak, 31 August 1989.

25. http://www.nsf.gov/nsb/documents/2000/nsb00215/nsb50/1970/mansfield.html

26. Letter from Senator John Glenn to Allan Cormack, 5 May 1982.

27. 'Results of a Survey of American Nobel Prize Winners Concerning Federal Funding for Research', Committee on Governmental Affairs, United States Senate, Washington, DC, July 1982.

28. Letter from Allan Cormack to Senator John Glenn, 25 May 1982.

29. Letter from Senator John Glenn to Allan Cormack, 30 July 1982.

30. Griffiths PA, Singer IM, Cormack AM, Hauptman HA, Weinberg S, 'Mathematics: the unifying thread in science', *Notices of the American Mathematical Society*, 33: 716–733, 1986.

31. Letter from Allan Cormack to Russell Sebring, 15 August 1986.

32. Letter from Allan Cormack to John Wanklyn, 1 February 1996.

33. E-mail correspondence with Jean Cormack, 15 May 2005.

34. E-mail correspondence with Michael Goitein, 9 April 2005.

35. Letter from Frank Jackson to Allan Cormack, 4 May 1981.

36. 'Go with the hunches', *Newsletter of the Alberta Heritage Foundation for Medical Research*, Edmonton, Canada, Winter 1981.

37. Note from Allan Cormack to Faculty Awards Committee, Account Number 527316, March 1980; and e-mail from Allan Cormack to David Weaver regarding Physics Discretionary Account 418123, 9 September 1992.

38. 'Nobel prize winners honored at the museum', *Newsletter of the Museum of Science and Industry*, Chicago, Illinois, July-August 1980.

39. Stigler SM, 'The odds of retirement', *Science*, 263: 1670–1671, 25 March 1994.

40. Letter from Allan Cormack to Stephen Stigler, 6 April 1994.

41. Letter from Stephen Stigler to Allan Cormack, 17 April 1994.

42. Cormack AM, 'Some family history', unpublished notes, Winchester, MA, page 3, 1996.

43. Letter from Allan Cormack to Amy and Len Read, 7 December 1997.

44. Letter from Allan Cormack to Amy and Len Read, 25 January 1992.

45. Interview with Margaret Cormack, 28 April 2006.

46. Cooper JS, Mukherji SK, Toledano AY, Beldon C, Schmalfuss IM, Amdur R, Sailer S, Loevner LA, Kousouboris P, Ang KK, Cormack J, Sicks J, 'An evaluation of the variability of tumor-shape definition derived by experienced observers from CT images of supraglottic carcinomas (ACRIN protocol 6658)', *International Journal of Radiation Oncology Biology and Physics*, 67(4): 972–975, 2007.

47. Interview with Robert Cormack, 23 April 2006.

48. Cormack RA, 'Beam profiles for x-ray rotation therapy', *Medical Physics*, 25(6): 879–884, 1998.

49. Letter from Allan Cormack to John Juritz, 20 September 1991.

50. Interview with Robert Cormack, 23 April 2006.

51. Interview with Godfrey Stafford, 15 October 2004.

52. Statement by Dr Barry Dorn, orthopaedic surgeon, regarding his patient Allan Cormack, 22 December 1994.

53. Programme for Electron Centennial Meeting, Cavendish Laboratory, Cambridge University, 21 March 1997.

54. E-mail letter from Allan Cormack to Archie Howe, 18 March 1997.

55. Interview with Margaret Cormack, 28 April 2006.

56. E-mail letter from Allan Cormack to John Wanklyn, 1 March 1998.

57. McLennan S, 'Memorial Service for Allan Cormack', Eulogy delivered at Goddard Chapel, Tufts University, 30 June 1998.

Figures

The author has attempted to trace the copyright holder of all the figures reproduced in this book and apologizes to copyright holders if permission to publish in this form has not been obtained.

Number	Page	Copyright	Caption
1.1	2	Christopher Vaughan	Map of Scotland
1.2	5	Christopher Vaughan	Allan Cormack's family tree
1.3	7	Cormack Family	Allan Cormack's great-grandparents: George and Margaret Cormack (left); and Allan MacLeod (right)
1.4	8	Cormack Family	Allan Cormack's grandparents: William Cormack (left); and Thomasina and Allan MacLeod (right)
1.5	9	Cormack Family	Allan Cormack's parents: George Cormack (left); and Amelia MacLeod (right)
1.6	11	Christopher Vaughan	Map of South Africa at the time of Union in 1910
1.7	14	Cormack Family	The Cormacks enjoy a family outing on the shore at John O'Groats, 1926
1.8	16	Cormack Family	Family portrait of Bill, Allan and Amy Cormack, 1930
1.9	18	Cormack Family	George, Allan, Jock and Amy on the road from Port Elizabeth to Uitenhage, 1934
1.10	20	Cormack Family	Berlin in 1938: The Reichstag (left) and Olympic Stadium (right)
1.11	21	Cormack Family	Allan and his mother Amelia at Nairn in Scotland, 1938

Number	Page	Copyright	Caption
2.1	25	Christopher Vaughan	On the slopes of Table Mountain
2.2	27	Rondebosch Boys' High School	AA Jayes was an inspiring teacher at RBHS
2.3	30	Cormack Family	Allan Cormack (centre) as secretary of the Debating Society with the two masters in charge: Wally Mears (far right) and Jack Kent (far left), 1941
2.4	32	Cormack Family	The cast of *Dear Brutus*, with Allan Cormack fourth from left, June 1941
2.5	34	University of Cape Town	University of Cape Town on the slopes of Table Mountain, 1945
2.6	42	Cormack Family	Cosmic ray experiment conducted by Godfrey Stafford as part of his MSc thesis; John Juritz and Allan Cormack assisted, 1946
2.7	44	Cormack Family	*Procrastination Crag* (left) and Allan Cormack climbing on the cliff face (right), 1943
2.8	49	Michael de Lisle	Cecilia de Lisle, Michael de Lisle, Ez Coaton, Allan Cormack, Chrystal Komlosy and Walter Powrie struggle with the donkeys, Anthony and Cleopatra, December 1945
2.9	50	Cormack Family	Wolfberg Arch (left) and a Bushman painting of an elephant hunt (right), 1945
2.10	51	Cormack Family	The mountaineers and their float with the theme 'Come landlord, fill the flowing bowl' participate in the Rag procession, March 1946
3.1	55	Cormack Family	St John's College chapel tower seen from the Backs across the River Cam
3.2	56	Cormack Family	Amelia, Allan and Amy Cormack at Cape Town harbour just before Allan boarded the *Durban Castle* for his passage to Cambridge, September 1947
3.3	65	Cavendish Laboratory	The members of the Nuclear Nit-Wits Club, 1948-49. Top: Allan Cormack, Godfrey Stafford and George Lindsey. Bottom: Charlie Barnes and Joan Freeman
3.4	68	Cormack Family	Dusty Miller and John Wanklyn on Ben Nevis, Christmas Day, 1947
3.5	77	Cormack Family	Barbara Seavey, an American student in Cambridge, 1949
3.6	79	Cormack Family	Allan Cormack, punting on the River Cam, 1949
3.7	81	Cormack Family	Louise Moore, Natalie Day, Barbara Seavey and Allan Cormack returning from France, April 1949

Number	Page	Copyright	Caption
3.8	82	Cormack Family	May Ball 1949. Back row: Allan Cormack and Peter Sturrock; Middle row: Frank Mercer and Ralph Bick; Front row: Elizabeth Poyser, Barbara Seavey, Dzunia Bick
3.9	84	Cavendish Laboratory	The Cavendish HT1 Cockcroft-Walton linear accelerator, 1949
3.10	87	Cavendish Laboratory	The Cavendish's 'modest cyclotron', used by Allan Cormack for his research on helium-six, 1949
4.1	92	Cormack Family	Mr William Foster, Mrs Amelia Cormack, Barbara and Allan Cormack (left to right) at Cape Town docks, February 1950
4.2	97	University of Cape Town	Professor RW James and Sir Lawrence Bragg at the University of Cape Town, July 1952
4.3	102	Cormack Family	Prospecting in Namaqualand with Stanley Amdurer, 1953
4.4	103	Cormack Family	Barbara and Allan at the summit of Waaihoek Mountain, 1952
4.5	104	Cormack Family	The Zeebasberg (left) with Mike Mamacos scaling the Gingerbread Face (right), 1952
4.6	106	Cormack Family	A Cormack family reunion, December 1955. Back, left to right: Allan; his mother Amelia; his brother Bill and wife Marjorie; William Foster. Front, left to right: Ian Cormack; Barbara and daughter Margaret; Amy and Len Read
5.1	114	Harvard University	The Harvard University Cyclotron with Lee Davenport (left) and Norman Ramsey (right), 1949
5.2	117	Tufts University	Allan Cormack's sketches of a wood and aluminium absorber (top), and a Geiger counter plus lead collimators (bottom), 1957
5.3	122	Tufts University	Allan Cormack's sketch of a phantom made from Lucite and aluminium, 1963
5.4	124	Christopher Vaughan	A replica of Allan Cormack's original turntable CAT scanner manufactured by the author, 2005. The original is in the Smithsonian Museum, Washington, DC
6.1	134	Cormack Family	David Ayres with the Čerenkov counter (left) and the 9-metre beam path where pions decayed into muons (right), 1966
6.2	135	Cormack Family	The Grand Canyon: Jean and Margaret in front, Barbara at rear, spring of 1967
6.3	137	Tufts University	Allan Cormack, the new department chair of physics at Tufts, 1968
6.4	139	Michael Goitein	Michael Goitein's algorithm successfully reconstructs Allan's phantom, 1971

Number	Page	Copyright	Caption
6.5	140	Institute of Physics	An object O contains positron annihilating material. An annular gamma-ray detector D defines a slice S of the object. The problem is to infer the density of the material from a number of different slices
6.6	141	Nobel Foundation	Hounsfield's original CAT scanner, released to the public in 1972
7.1	157	National Library of Medicine	Panel of pioneers (left to right): William Oldendorf, Allan Cormack and Ron Bracewell listen to Giovanni Di Chiro reminisce about his experience with computer tomography, October 1976
7.2	167	Susan Lapides	Allan lectures to his freshman physics class on his line integral problem, October 1979
7.3	168	Cormack Family	Allan at a press conference, with Barbara and Jean seated on the couch, October 1979
7.4	169	Susan Lapides	Allan explains to the audience how his CAT scanner works, October 1979
7.5	175	Torgny Greitz	Allan Cormack and Godfrey Hounsfield enjoy a story told by Torgny Greitz, December 1979
7.6	176	Cormack Family	Tom Baker, who played the part of Dr Who, entertains Robert Cormack, December 1979
7.7	178	Torgny Greitz	Allan receives the Nobel Prize from the King of Sweden, 10 December 1979
7.8	179	Cormack Family	Allan delivers his short speech, on behalf of himself and Godfrey Hounsfield, 10 December 1979
8.1	182	Tufts University	Tufts University confers an honorary DSc degree on Allan Cormack, May 1980
8.2	184	South African Post Office	The South African Government commemorates its Nobel Laureates with a set of stamps, November 1996
8.3	189	Harvard University	Andreas Kohler, director of the Harvard Cyclotron Laboratory, and long-time collaborator with Allan Cormack
8.4	190	Tufts University	Allan holds a custom-designed beam shaper used in radiotherapy treatment
8.5	199	University of Würzburg	The centenary of Willem Röntgen's discovery of X-rays is celebrated in November, 1995
9.1	208	Tufts University	Allan's love of mathematics comes through clearly in a lecture, 1985
9.2	209	Museum of Science and Industry	Victor Danilov with 1979 American Nobel Laureates (left to right): Herbert Brown, Allan Cormack, Steven Weinberg and Sheldon Glashow, April 1980
9.3	212	Cormack Family	The Cormack family in Iceland: Barbara, Allan, Margaret, Robert and Jean, 1986

Number	Page	Copyright	Caption
B.1	225	Nobel Foundation	Beam of gamma rays of initial intensity I_0 passing through body along line L and emerging with intensity I
B.2	226	Nobel Foundation	X-ray reconstruction of a circularly symmetrical object in 1957
B.3	227	Nobel Foundation	Computed tomography scanner used in 1963; its cost was approximately \$100
B.4	228	Nobel Foundation	Results obtained with the 1963 CAT scanner. (a) Aluminium and Lucite model of a head with two tumours. (b) Solid line: actual values of the attenuation coefficient along the line OA. Dots: experimental results
B.5	231	Nobel Foundation	Number-distance curves for photons (X-rays) and protons penetrating a plastic
B.6	232	Nobel Foundation	Scans of a normal brain with (a) X-rays; and (b) protons
B.7	232	Nobel Foundation	Scans of a heart with myocardial infarction, fixed in Formalin, with (a) X-rays; and (b) protons
B.8	233	Nobel Foundation	Ionization as a function of distance for 150-MeV protons penetrating polystyrene
B.9	234	Nobel Foundation	The hatched region represents a plane intersecting a sphere
B.10	234	Nobel Foundation	The polar coordinates (p, ϕ) specify a circle passing through the origin

Index